AF306389

CONSCIOUSNESS, QUALIA AND ROBOT MINDS

Series on Machine Consciousness
ISSN: 2010-3158

Series Editor: Antonio Chella *(University of Palermo, Italy)*

Published

Series on Machine Consciousness – Vol. 6

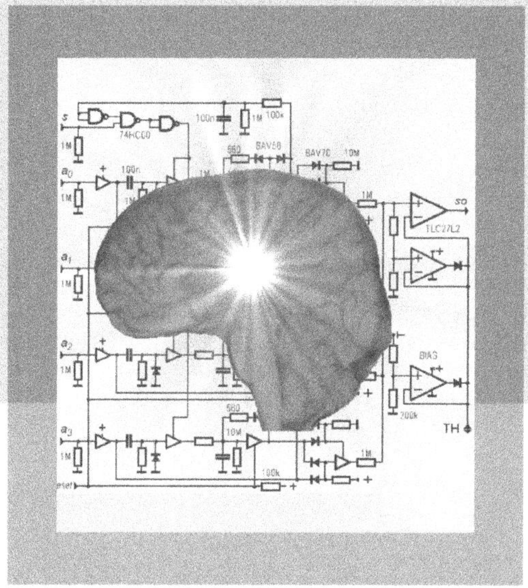

CONSCIOUSNESS, QUALIA AND ROBOT MINDS

Pentti O Haikonen

University of Illinois Springfield, USA

World Scientific

W JERSEY · LONDON · SINGAPORE · BEIJING · SHANGHAI · HONG KONG · TAIPEI · CHENNAI · TOKYO

Published by

World Scientific Publishing Co. Pte. Ltd.

5 Toh Tuck Link, Singapore 596224

USA office: 27 Warren Street, Suite 401-402, Hackensack, NJ 07601

UK office: 57 Shelton Street, Covent Garden, London WC2H 9HE

Library of Congress Cataloging-in-Publication Data

Names: Haikonen, Pentti O. author

Title: Consciousness, qualia and robot minds /
    Pentti O Haikonen, University of Illinois Springfield, USA.

Description: New Jersey : World Scientific, [2026] | Series: Series on machine consciousness,
    2010-3158 ; vol. 6 | Includes bibliographical references and index.

Identifiers: LCCN 2026004154 | ISBN 9789819827824 hardcover |
    ISBN 9789819827831 ebook for institutions | ISBN 9789819827848 ebook for individuals

Subjects: LCSH: Artificial intelligence | Robots | Qualia | Consciousness

Classification: LCC Q335 .H345 2026

LC record available at https://lccn.loc.gov/2026004154

British Library Cataloguing-in-Publication Data

A catalogue record for this book is available from the British Library.

This book does not contain AI-generated text.

For any available supplementary material, please visit
https://www.worldscientific.com/worldscibooks/10.1142/14720#t=suppl

Desk Editors: Eshak Nabi Akbar Ali/Veronica Lee

Typeset by Stallion Press
Email: enquiries@stallionpress.com

To Humanity.

Preface

Is consciousness immaterial? The age-old observation is that the human body consists of matter, while the conscious mind appears to be immaterial. This belief has led to the philosophical mind–body problem about the interaction between the material body and the immaterial mind: How can our immaterial thoughts and imaginations control our physical actions? How can any immaterial, nonphysical substance affect physical matter? What is this immaterial mind actually?

Matter is matter, but immateriality is a mystery. The mind–body problem is thousands of years old, and in the course of time many prominent philosophers have tried to solve it, in vain. The situation has still been the same. A quick search through the Internet reveals the common opinion: Nobody has been able to truly solve the mind–body problem, and there is a distinct possibility that the mind–body problem will always remain unsolved; immateriality and the mind–body problem may well be beyond the grasp of the human mind. If this were really so, then also the phenomenon of consciousness could not be explained.

It is obvious that consciousness cannot be explained without explaining first the mystery of immateriality. This is seen as a serious problem that has major implications to the studies on phenomenal consciousness, the enigma of qualia and the development of conscious robots. The research on machine consciousness will remain fruitless, if the mystery of immateriality and the mind–body problem are not solved.

This book explains immateriality, dissolves the mind–body problem and explains what consciousness really is. These explanations are simple, maybe surprising, yet obvious and technically sound; no exotic theories, substances or "new physics" are involved. These explanations and solutions are not only philosophical theories; they are practical and technically applicable, as is shown in the chapters of this book. They have implications to the philosophy of mind and to the engineering quest for conscious machines. It is seen that there are no fundamental obstacles to the development of true conscious robots.

About the Author

Pentti O. Haikonen, D.Sc. (technology) is an adjunct professor at the Department of Philosophy, University of Illinois at Springfield. Formerly he was the Principal Scientist on cognitive technology at Nokia Research Center. He has a wide professional experience in electronics and information technology, and he is a well-known contributor to the field of machine consciousness with several books and many journal papers.

Pentti O. Haikonen is the author of *Existence, Origin and Weird Technology* (World Scientific, 2023), *Consciousness and Robot Sentience* (World Scientific 2012, second ed., 2019), *Robot Brains* (Wiley, 2007) and *The Cognitive Approach to Conscious Machines* (Imprint Academic, 2003).

Acknowledgments

I want to thank the publisher for the possibility to write this book and the good people at World Scientific for their kind help and expert efforts.

My sincere thanks go to Professor Antonio Chella for his kind support and the possibility to write books for the World Scientific Series on Machine Consciousness. Grazie mille!

Muchisimas gracias also to PhD Raúl Arrabales Moreno, who has kindly presented my work on his excellent websites.

I also want to thank my older brother D.Sc. Terho (Max) Haikonen for the inspiring discussions on scientific issues.

I want to thank my Media Artist son Pete Haikonen for pleasant cooperation and the artistic illustrations found in this book and also for the various demo videos about my robot.

My truly heartfelt thanks go to my wonderful wife and soulmate Sinikka for her continuing devotion, encouragement, support and patience.

Finally, I want to thank you, my respected reader, for your much appreciated interest in my modest work.

Pentti O. Haikonen
January 1, 2026

Contents

Chapter 1

Introduction

1.1. Artificial Intelligence and Consciousness

Are chatbots conscious entities?

According to Science Fiction, Artificial Intelligence (AI) will eventually exceed human intelligence. Considering the fast advancements in AI research, AI is now foreseen to achieve consciousness in no time at all, and there will be robots with conscious minds. But there may be some fundamental and overlooked issues still to be seriously considered, see e.g. [Bołtuć, 2020].

Artificial Intelligence is already here. It is in our smart phones, cars and appliances. It is in our wellness rings that monitor our health and tell us that we did not sleep well last night. It is also in the chatbots, computer programs that converse with humans.

Chatbots are the ingenious AI assistants that we encounter when we try to contact banks and businesses. Chatbots seem to understand natural languages and also seem to be able to answer questions via their own reasoning. Chatbot assistants are already so clever that on occasions, human intelligence seems to be no match for their logic.

Already in the last century it was predicted that AI will become more and more intelligent. It will become Artificial General Intelligence (AGI), and eventually its intelligence will surpass the human intelligence. This would happen very soon, and moreover, AGI will then also become conscious; that moment will be the "AI singularity". In fact, it has been recently claimed in public that certain generative chatbots were already not

1

only intelligent, but also conscious. If they were, it would be the glorious pinnacle of AI, and perhaps the ultimate achievement in high technology, with profound implications to the philosophy of mind, existentialism and metaphysics. This technology might also soon allow conscious Internet and the creation of true conscious robots. Recently advertised advances in chatbot technology are seen to confirm this.

Chatbots appear to be experts on all matters. Yet in reality, they are only search machines with extensive knowledge bases and cosmetically engineered interfaces, which produce the illusion of an actual person behind the interface.

Chatbot knowledge bases are created by gathering information available in the web. In this way chatbots might become immensely knowledgeable. However, in addition to facts, the gathered information includes fiction, propaganda and straightforward insanity. This will lead to the contamination of knowledge base, and there is not much that the current AI can do about it due to AI's fundamental limitation: AI does not operate with meanings, and therefore it does not actually understand anything and cannot have humanlike intelligence.

More computing speed, more memory, more code command lines. That should solve the limitations of current AI, and true intelligence and consciousness should be soon achievable in this way. But unfortunately, more of the same does not lead to any radical change in the essence, it only leads to more of the same. Pyramids are still only piles of stones. Large ones, yes, but still piles of stones, regardless of any hype.

Media hype has led to the popular misconception about AI: AI is easily taken as a real entity with the humanlike ability to think, reason and produce its own thoughts. But no amount of hyping makes false claims of performance true and real, and current AI is not by far what it is hyped to be. The inherent limitations of current AI have not been overcome, and the claims of consciousness in chatbots have not been taken seriously by the scientific community for good reasons.

The fundamental limitation of current AI originates from its way of operation; it is based on algorithms that are run in computers as symbolic rule-based computation. This computation handles symbols without meanings; actual meanings are neither imported nor utilized.

Computers accept information only in the digital form that is, binary data. This data is symbolic, and as such it does not carry explicit meanings. Symbols have to be explained, but how to explain them, if there is nothing else than symbols. This problem is known as *the symbol grounding problem.*

The symbol grounding problem is not solvable in pure symbol processing systems. That is unfortunate, as computers are pure symbol processing systems accepting only symbols as their inputs. This has a consequence; current computational AI can manipulate only symbols, not their meanings. In contrast, human thinking operates with meanings.

Consequently, AI programs neither think nor understand anything, they just produce what is programmed. This limitation is recognized well by competent AI researchers, and therefore current AI is seen as weak AI; its intelligence is just an illusion. The limitation has to be overcome and other approaches have to be tried. Accordingly, research is done on Artificial General Intelligence (AGI) and Strong Artificial Intelligence (Strong AI), and it is hoped and hyped that this research would lead to true humanlike machine intelligence and possibly to machines and robots with conscious minds.

However, there are challenging problems and difficult issues to be noted. The nascent Strong Artificial Intelligence does not appear to have what it takes to become conscious, but would it be able to achieve consciousness in the future? May be not. It is quite possible that it is a dead-end approach. In that case what would have to be done differently in the quest for machine consciousness? In order to solve these issues, the mysteries of self, mind and consciousness must be studied, and the recognized problems have to be analyzed and solved.

1.2. The Mysteries of Conscious Minds and Qualia

Entering the forbidden waters of knowledge.

The self, mind, and consciousness are humanity's ultimate mysteries that have been seen to involve many difficult issues. It has even been proposed that these might even be unsolvable by science's methods. What is self, the one who observes? What is mind, what is its relation to self and consciousness, and how does it control the material body? What would it take to build machines with minds and consciousness?

Scientific research has produced our high technology and advanced research instruments. Finally, it should be possible to find concrete scientific explanations for these mysteries; this should be worthwhile challenge to ambitious AI researchers and robotics engineers.

Yet, even rather recently in the 1980's, the phenomenon of consciousness was still seen to be beyond the reach of true science and rather belonging to the realms of pseudoscience and Science Fiction. Thus, consciousness was not fit for serious scientific research and was not to be approached by respectable engineers. Physics and engineering sciences are about what can be observed, measured and treated with verifiable, working mathematical models. Consciousness has not been seen as this kind of a phenomenon, and therefore clear-minded scientists or engineers were assumed to stay away from it, in order to retain their prestige, careers and sanity.

However, history shows that the advancement of science has involved bold approaches to phenomena that initially have not satisfied these conditions. This is and has been a necessity; we will not learn to understand the unknown, if we are not allowed to approach it. This applies to the mystery of consciousness, too. Mere philosophical approaches are not sufficient here by far; the problem of consciousness should be approached from biological and engineering points of view and with insights from psychology and cognitive neuroscience.

Mere theoretical research and observations are not always enough. The proven old engineering wisdom has it that we can understand a phenomenon properly only if we can create it artificially. Work teaches the worker, and hands-on experiments will provide new insights. The problem of consciousness is not different; its satisfactory explanation cannot be merely a philosophical or theoretical one. The explanation of consciousness must be verified by actual artificial realizations, like conscious machines and robots. The proof is in the doing, and knowledge is brought by experimenting.

The designing an artifact with conscious mind faces hard challenges. What is mind, and what would be its relation to self and consciousness? The ancient view has been that mind and consciousness are supernatural and immaterial, while the body is a material entity. If this were the case, a conscious machine could not be built using matter and material processes only. The supernatural mind would remain unexplainable and unattainable, and the quest for conscious robots would be futile.

Humans are individual beings with brains, conscious minds and mental experiences. The brain is a physical, material entity, and thus at least in principle, its operation could be solved by external inspection. This kind of research is being done, and the brain's basic structure as a neural network is being observed in more and more detail. Neural activity patterns have been observed and their correlations with the reported mental content have been noted.

But there is a problem. Detected neural activity patterns are only presentations of the internal brain activities, created by the brain imaging instruments. No mental impressions are detected as they are because our existing instruments cannot access them.

In contrast, our personal notion is that we observe our sensory percepts and imaginations directly as such, without even a hint of any underlying material neural machinery or activity. This leads easily to the conclusion that the mind is immaterial, while our body is definitely material. Ghosts and spooks (if they existed) would be immaterial; what kind of entity would an immaterial mind be? That is a mystery.

The idea of a material body and an immaterial mind is age-old. In the philosophy of mind, the hypothesis of the material body and the immaterial mind as two different substances is known as dualism.

Dualism is supposed to explain how things are, but it has a serious unsolved problem, known as the mind–body problem. This problem relates to the interaction between the material body and the immaterial mind; how to dovetail the existent with the materially non-existent. The matter (including all physical entities) exists, while the rest is nonphysical and supernatural, without any scientific proof of its existence. This produces the *mind–body interaction problem*, and philosophers have worked hard to circumvent it by inventing properties for the materially non-existent (property dualism etc.). All in vain.

According to modern physics, supernatural substances do not exist, and therefore dualism should be rejected. What exists and is observable cannot be explained by the existence of the non-existent and non-observable. But alas, the rejection of dualism will not remove the mind–body problem. Unfortunately, this problem is a fundamentally relevant issue also in modern machine consciousness research. The mind still appears to be immaterial, and so should the mind of any conscious robot.

The apparent immateriality of the mind is related to the mystery of qualia, the internal qualitative appearances of our conscious mental content. How can qualia appear as they do in an apparently immaterial way, and how does the material brain produce apparently immaterial qualitative impressions? This is known as *the hard problem of consciousness* [Chalmers, 1995]. It can be seen that the problem of dualism will vanish, as soon as the mystery of qualia is solved.

Qualia are related also to the *symbol grounding problem*. There is no symbol grounding problem in humans, as our senses produce direct and self-explanatory information. Seen colors are colors, heard sounds are sounds and so on. These and all the other sensory experiences are direct, and do not require any explanations. Qualia solve the symbol grounding problem by providing pieces of outside information to be used as such, or as symbols associated with other pieces of information. Qualia would do the same in machines if they were able to utilize them. But another problem remains; how do neural signals convey meaning?

Humans interact with their environment. For this, humans have senses and muscles. Senses provide information about the environment, and muscles execute the physical interaction. Perception is direct and allows seamless interaction. We can reach out for things and grasp them without any explicit calculations. Likewise, we can execute our imagined actions and we can imitate seen actions. Imagined actions are mental and apparently immaterial, while physical actions are material.

Here the old mind–body interaction problem arises in practical context. How can imagined acts be transformed into material neural commands to the proper muscle neurons with locations that are unknown to the imagination process? This is a real and tricky issue to be solved.

In this book new solutions to these issues and problems are searched and presented from philosophical, technical and neural points of view, especially in the context of qualia, consciousness and mind–body interactions in robots with minds; the mysteries of the self, mind and body.

Chapter 2

The Self and the Mind

2.1. I Am Me

Sentio, ergo existo — I feel, thus I am. If you doubt your existence, go to the dentist.

Infants will slowly become self-aware. They will find their bodies, and will learn how to execute physical motions and actions at will and with increasing dexterity. By moving around, they will note that there is a difference between them and their environment. Around the age of two years, babies will begin to recognize their mirror image. At about the same time they will learn to want and own things that they see, and as soon as they learn to speak a little, they will use that skill to express their will. They have developed a rudimentary sense of self and own will.

I am the one who is behind my eyes. Herbert Spiegelberg [1961] noted in the sixties that a certain personal experience has been neglected by both philosophy and psychology, namely the "I am me" experience. That is the moment of sudden awakening, when a child realizes in a flash that she/he exists, and exists as an individual, separate from others.[1]

---

[1]When I was four years old or so, I had this kind of "I am me" experience. I suddenly realized that there was something special in my existence. I was me, but wherefrom did I become to be inside me? I felt that if I only could formulate this question properly, I would soon find the answer, and the existence of my brother and my sisters would have some part there. I felt that the answer was right there within my reach, but just then in a flash of dizziness it was all gone. It was as if I had tried to enter a forbidden area of knowledge. I had this experience again once or twice, with the same outcome.

7

The "I am me" or the "I am inside my body" experience is not only about being as an individual, it has also an existential component; the realization that one exists now, but has not been here forever. For the child, the world around and the parents are given, and have always existed, while the child has not. Therefore, children will ask: "Where did I come from, why now" and "Why am I me, why not one of my siblings or one of the neighbor's kids". These contemplations may be accompanied by a moment of fleeting dizziness.

Technically, I am me because I have always been me. I have biologically existed from the time when first living cells emerged, but not as a conscious entity until now. The first living cell began to divide resulting in a chain of cell–cell–cell, then egg cell–egg cell–egg cell and so on. In the course of this unbroken chain, DNA grew and accumulated mutations, resulting in lineages of various living beings. As a male I am the endpoint of one of these unbroken lineages, and my personal lineage could, in principle, be traced back to the first living cell eons ago. The starting point of my personal lineage is thus the starting point of my identity.

Identity is not the same as "self". Wikipedia explains that according to philosophy, the self is an individual's own being, knowledge, and values, and the relationship between these attributes. APA Dictionary of Psychology explains that the self is the totality of the individual, consisting of all characteristic attributes, conscious and unconscious, mental and physical. These and other more or less similar definitions of self have one thing common: They are vague and very much useless.

In common usage "self" is a word that refers to one's own body and person. The words I, me and myself refer to the person who is using them. (Note that "self" can also be used to indicate the selfhood of others; "he himself did it").

Little children may not learn the usage of the concept of self at once. They may not use the word "I" to refer to themselves. Instead, they may use their name in the same way as their parents do. They may say, "Jack wants candy" instead of saying "I want candy".

My body is my body not because I can see it, but because I can feel it. The numb hand experience proves this. It may happen that at night one wakes up with a numb hand. This may lead to the scary experience of a foreign hand; when the numb hand is touched by the awake hand, the numb had will not respond, and consequently will not be recognized as one's own.

Having hands to touch oneself is an important contributor to the self-concept. Touching one's own body will lead to the experience of ownership and unity, because two simultaneous touch sensations will follow; the one at the touching hand and the other at the touched body part. Likewise, if I feel pain, the hurting body part is mine. The visual recognition of one's body comes after these experiences.

Mirrors allow one to see how one looks, and this contributes to one's mental self-image. A mirror is a marvelous thing; people know that it inverts its image in the horizontal direction, but not in the vertical direction. This obvious function is impossible to explain properly, even though a good number of ridiculous explanations have been presented, also by AI. There cannot be an explanation for this mirror image inversion because there is no such thing in the first place; it is just a misperception. In the mirror image your left hand is on your left side, and your right hand is on your right side, nothing is inverted. This helps the recognition of one's mirror image.

Little children will learn to recognize themselves in the mirror at around the age of two years. Mirror image recognition is not a difficult task, and can be learned if, for instance, the baby has learned the control of hands at will. Now, for example, if the left hand in the mirror raises when the baby raises her/his own left hand, and if the hands in the mirror move at the will of the baby, then the mirror image belongs to the baby, and the mirror image is the baby's image. (This is rather similar to a child's first observations of her/his own shadow. The shadow follows, and may not be necessarily recognized at first as what it is, and the child may be scared of it.)

Mirror image self-recognition will help in the developing of the mental self-image of the child's appearance; eventually the child will become self-aware of her/his looks, also in comparison with others' looks. The inability to recognize one's image in a mirror may be a symptom of depersonalization-derealization disorder. Mirror self-image recognition in robots has been investigated by Takeno *et al.* [2005] among others, and also the author, see the Chapter 24.

Mental self-image is more than the recognition of one's body. It is also about how one finds oneself as a person; how one perceives one's position in respect to others and what kind of person one is. Mental self-image is related to memories of personal history.

Mental self-image can be good or bad, and it affects one's self-esteem and self-confidence. Distorted mental self-image may lead to bad behavior and mental problems.

2.2.   What Is the Self Inside?

From a pragmatic point of view, it is obvious that a person's concept of self arises from the person's observations of the person's own breathing, actions, thoughts, feelings and the body with its sensations. This view is shared by others. For example, Antonio Damasio has presented that the basic sense of self is grounded to the percepts of the body [Damasio, 2000, p. 22]. Apparently, this is so, but a problem remains: Who is the perceiver or observer? The apparent answer would be: The self itself is the observer. But what is an observer?

My body is my body and I can observe it. I find myself being the observer inside who perceives whatever my senses deliver to me. My thoughts are my thoughts and I can perceive them.

I observe myself observing myself. The perception of oneself as a separate entity enables the impression of being the perceiver self. Observing oneself observing oneself is the *self-loop* that generates the experience and concept of self.

Here the observer oneself observes oneself, but who is the observer, actually? This question leads to *the observer problem*. Obviously, the self is somehow the observer, but on the other hand, the self is only a concept. Concepts are not entities that do something, therefore the self cannot be taken as an observing entity without further elaborations.

The observer problem is a truly fundamental question that is related to the concepts of mind and consciousness and also to the perception of inner mental content. A solution to the observer problem is presented later on in Chapter 4, and it may not be what you might expect.

2.3.   What Is Mind?

What is mind? That should be the first question to be considered by every serious student of psychology and nowadays also by every wannabe designer of conscious AI and sentient machines with minds.

The question about the essence of the mind should be an easy question to answer. After all, don't we all have minds, and don't we know what "mind" means. According to folk psychology, mind is the very entity that receives information from the senses. Mind thinks, remembers, has opinions and controls us. Mind's physical location is usually felt to be in the head, behind one's eyes and between the ears, and it is known that the brain is the organ where the mind resides. The mind is understood to be the one who is able to imagine things: "I see it with my mind's eye". The self is seen as the owner of its mind, as we can say: "In my mind, things are in this way".

In theology, the soul and the mind are seen to be related, but are not necessarily considered as the same entity. In many religions soul is the spiritual entity that after the death is supposed to go to heaven (or, depending on circumstances, to the other place, where the welcome is ever so warm). It is not clear, whether the soul would take the mind with its personal memories along.

Folk psychology and theology aside, sciences and philosophy do not have any universally agreed definition for mind. Actually, "mind" is just a name for a vague ill-defined concept of something that might be an entity or a process or a collection of functions or even all of these.

What is in a name? The inventing of a name is not the same as explaining something, and an invented name is not a proof that something exists. Naming easily suggests that there is a real, definite entity behind it. But naming does not make non-existent entities real. In science, it must be known what is being named — investigation, analysis and definition must come first, only then the name. Nevertheless, the concept of mind is used in the philosophy of mind, no matter how vaguely it is defined, and it is used also here.

There are two fundamental mysteries of the mind. The first mystery is the observer problem. The second mystery is the immateriality problem: The mind appears to be immaterial, while the brain is definitely material.

The material brain is supposed to be the home of the mind, but our introspection does not reveal any material processes taking place when we are thinking and imagining things. Therefore, our mind appears to be some immaterial entity. In contrast, when we inspect our body, we will soon notice that our body consists of matter. It has weight and it takes space, it is a physical object. Therefore, it would be easy to conclude that the mind and the body are of two different and separate substances that are a material substance and a non-material or immaterial substance. There would be a spiritual self and a material self.

This mind–body issue has profound implications on machine consciousness research; if the conscious mind were truly immaterial and non-physical, it could not be reproduced by material and physical means including quantum physics and electromagnetic fields, and consequently the pursuing of conscious machines would be futile.

However, it is presented later on in this book in that regardless of the mind's immaterial appearances the mind is not truly immaterial; the immateriality of the mind is only an impression. Thus, conscious machines may be possible after all, but their construction may not be the kind of straightforward programming that contemporary AI people seem to assume. And they should have already noticed that.

Wannabe designers of conscious AI and sentient machines should make themselves aware of the essential issues of the self, mind, the mind–body problem and qualia.

Chapter 3

The Mind–Body Problem

3.1. Much Ado About Nothing and Everything

How does the immaterial mind control the material body? Why is this question so difficult that nobody has been able to answer to it properly?

We have a material body. When we execute physical work, we can easily note that the work is done by our material muscles; our interaction with the material world is physical and often hard. But our mind seems to be different and operating without any material processes. When we look around and see the things out there so effortlessly, we do not see any matter flowing between the seen object and our eyes; the interaction must be immaterial. Likewise, when we think, we are not aware of any physical work or material processes taking place in our brain. Our mental contents, like thoughts and imaginations, appear to be immaterial. But how to explain these observations?

The early explanation was that mind and body were made of two different substances. The body was a material entity while the mind was an immaterial spiritual entity, the soul. An apparent additional proof for this was seen easily. When a person dies, the lifeless body looks the same for a while. No matter has been lost, but something must have left the body, as it does not function anymore. This something would be the immaterial soul or the similarly immaterial élan vital, force of life.

The original mind–body theory was formulated by early philosophers, to whom matter was very concrete and touchable. All things around them consisted of matter and operated as material and mechanical contraptions,

13

simple machines. The human body consisted of matter, and everything the body did, was a material action. But in contrast, the mind and its contents were seen to be immaterial. Later on, this concept became to be known in the philosophy of mind as *mind–body dualism* and also as *substance dualism*, the hypothesis of the coexistence of two different substances; material and immaterial substances.

The concept of dualism is an old one, but it became universally known by the works of the French philosopher and mathematician René Descartes (1596–1650). Descartes' mind–body dualism is also known as Cartesian dualism.

At first sight mind–body dualism appears to explain neatly the human being as the union of the immaterial mind and the material body. However, a closer inspection reveals that dualism comes with an insurmountable problem; the problem of mind–body interaction. Physical matter and a non-physical, metaphysical substance do not easily interact, (especially if one of these is non-existent). Matter is easily observable, and its amount and properties can be measured by material instruments. On the other hand, immaterial substances cannot be observed or measured by material instruments. Matter is matter, but what would be the immaterial substance, and how would the immaterial mind interact with the material body? This is a tough question.

In the philosophy of mind this issue is known as the mind–body problem of dualism: In the material world the immaterial is nothing, and in the immaterial world matter is nothing. So, how can they interact and work together? In effect, mind–body dualism has three problems that are: (1) The information acquisition problem, (2) The interaction problem, (3) The internal appearance problem.

The information acquisition problem relates to the fact that the observed world is material, and our senses, like eyes and ears, are also material units and produce material responses. What good would material senses and their material responses be for the immaterial mind?

Obviously, an immaterial mind would not be able to acquire information about material happenings on its own. But how would the use of material senses change anything, as they would deliver the acquired information in material form? And on the other hand, if the immaterial mind were able to acquire information about material happenings on its own, what would material senses be need for?

The interaction problem relates to the apparent impossibility of material interactions between the immaterial mind and the material body; how could the immaterial mind control the material body and its physical actions, and vice versa?

The internal appearance problem relates to the way in which the conscious content of the mind appears internally with its qualities, in a phenomenal and immaterial way. In fact, this problem is the root of all evils here, and it should be solved at first. If it turned out that the immaterial appearance of the mental content resulted from physical processes, then the immaterial substance could and should be rejected, and consequently the information acquisition problem and the interaction problem would disappear. That would be good, but unfortunately, this is has been the trickiest problem of them all to be solved. This is also a wider problem; all proposed solutions to the mind–body problem must solve the internal phenomenal and immaterial appearance problem, otherwise they have solved nothing.

Descartes understood that the interaction between the material and the immaterial was a serious problem, but no matter how hard he thought about it, he could not find any solution. Finally, he proposed that the interaction between the material and the immaterial would somehow take place in the pineal gland in the center of the brain, and ultimately, this interaction would be facilitated by God.

Many philosophers found Descartes' explanation unsatisfactory. God should not be included in any scientific explanations, as then you would have to prove first that God existed. If this road were taken, you would ultimately end up admitting that scientific explanations were not necessary at all for any physical phenomena, as ultimately, these phenomena would be acts of God. That would transform science into religion. Cartesian dualism had to be rejected.

In the terms of philosophy, the mind–body interaction problem can be removed by denying either the existence of the immaterial or by denying the existence of the material. In this way there will be no interaction to be explained. Both explanations have been proposed. The denial of the material world leads to the philosophical theory of *idealism*, and the denial of the immaterial leads to the philosophical theory of *materialism*.

3.2. Mind–Body Idealism

Mens sana non indiget corpore. Healthy mind does not need a body.

If indeed the mind were immaterial and consciousness were a property of an immaterial universe, then, for what purpose would the material world be needed? Bishop Berkeley (1685–1753) considered this question and concluded that there was no need for any material world at all, as immaterial minds do not need matter for anything. (See e.g. Stanford Encyclopedia of Philosophy). What we take as the material world including our lives, pleasures and sufferings, is not real. All this is only an illusion created by God. What we sense and believe to be properties of the world and our body are only immaterial impressions in our immaterial minds.

Berkeley's idea presents that the apparently material universe is only a kind of imagination of God. In the material sense we do not exist at all; this is just a weird game of illusions and apparitions. This is an excitingly wonderful solution to the mind–body problem, as it removes two problems with one blow; namely the interaction problem and the need to explain what matter is. Matter does not exist and everything is immaterial; there is no mind–body problem.

But there is a hefty fee to pay: How to explain the immaterial mind? Without this explanation, idealism would be uselessly incomplete. Matter has been explained away, and has been replaced by immaterial impressions and the realm of the supernatural. How do you explain this without explaining the existence of God and opera Dei? Should these be simply considered as given, and should natural sciences be replaced with faith and theology?

The rejection of matter did not and does not please everybody. After all, physical matter and phenomena are consistently observable by our senses and measurable by our instruments, while the supernatural is not. Nevertheless, Berkeley's idealism is still alive, nowadays in the modern forms of computer simulation theories, like that of Philosopher Nick Bostrom [2003].

Simulation theories maintain that all what we believe to be real, are not; they are just illusions created by simulation. In the real reality, we live in a computer simulation created by superior beings. The world around us with all its things is actually a simulation. You, my respected reader, are reading right now a simulated book, written by nobody. Your friends and close ones are only simulations, as well as all the other people; they do not exist, they are only illusions in your mind. There is only you, and in fact, even you would not actually exist. You and your mind would be just a simulation — and in fact the only simulation; that would make things so much easier. The simulation of one person's instantaneous mental content is far simpler than the simulation of the whole world and universe. And even more, the simulation of only one present moment with faked memories would suffice, as at each moment of time, the present is the only thing that we experience.

We live in a few minutes time loop. But then, who would be the superior simulator?

A good theory explains more with less; otherwise it would only increase the amount of things to be explained. Simulation theories are not good ones in that respect, and do not deserve much respect.

Simulation theories are just entertaining philosophical speculations, not science, and as such better to be left for Hollywood. Natural sciences are based on the assumption that there is a coherent material world and phenomena that can be consistently observed, measured and modeled by testable theories.

3.3. Property Dualism

Could matter have both material and immaterial properties?

Property dualism accepts that the material body and the apparently immaterial mind coexist. However, the mind is not genuinely immaterial, it is a product of or a companion to matter. Property dualism presents that material systems may have two different kinds of properties that are the physical properties and the apparently immaterial mental properties, the contents of the conscious mind. The mental properties arise from material processes, and appear on top of those [Chalmers, 1996]. The problem is seen to be solved, if the mental properties are taken on a par with the material properties.

In property dualism the existence of the immaterial is denied, and consequently the material body — immaterial mind interaction and the nature of the immaterial substance do not need to be explained. Physical properties are physical properties, and the mental properties are mental properties that emerge from physical properties. That is it, and the case is closed.

Unfortunately, that is not so at all. The denying of the immaterial opens up three hard problems.

Surprisingly perhaps, these new problems are not about the mystical emergence of the apparently immaterial properties; the problem is not about the properties at all. The real problems are the mental appearance and especially — the appearance to whom, the observer of the appearance. In the traditional mind–body theory the observer is the immaterial mind with its immaterial contents, but now the immaterial mind is denied, and consequently there seems to be nobody at home. The denial of the immaterial has removed the observer; a property as such is not an observer. This is the observer problem.

Could the material system in itself be the observer? Yes, in ways it could. There are material systems that can access their own internal states and operate with them. Feedback control systems and computers — and also mechanical calculators — are examples of these. But in that case, what would be the need for mental properties, then? No need, no mental properties, but then not only would the immaterial mind be denied, the non-immaterial mind would be rejected as well. It can well be argued that any robot utilizing this approach would be a mindless zombie.

Property dualism does not explain how the physical processes could produce apparently immaterial impressions in the mind. Property dualism may solve the problem of two substances, but in doing so, it appears to produce three additional problems, namely the problem of the emergence of matter's apparently immaterial properties, the apparently immaterial appearance of mental contents and the problem of the observer. That is an uncomfortable situation, and we are back at the square one.

3.4. Mind–Body Materialism

Is consciousness really matter?

Mind–body materialism is based on a simple assumption: There are no ghosts. Only matter and material interactions exist, and non-matter, supernatural substances and phenomena do not exist. This means that the mind cannot be immaterial.

A hardcore materialist may be surprised by the fact that modern physics does not deny the existence of non-matter. On the contrary, both matter and non-matter do exist. Unfortunately, this does not save dualism or idealism; the supernatural is still denied.

In physics, matter is any substance that has mass; that is, it has weight and inertia, and it can have kinetic energy. Protons, neutrons, electrons, atoms and molecules are examples of matter. Material particles can collide and exchange kinetic energy with each other. They can also interact with each other in other ways in nuclear and chemical reactions.

The physical non-matter does not have these properties. Examples of physical non-matter are photons, electromagnetic fields and waves. These do not have mass, weight, inertia or kinetic energy. They do have or convey radiation energy. Photons do not interact with each other in normal (linear) conditions. At very high energy levels the situation is different. For instance, the collision of two very high-energy gamma photons may produce a material electron-positron pair.

The interaction between the physical non-matter and matter takes normally place via charged particles, such as electrons. Electromagnetic fields have electric forces that can accelerate charged mass particles like electrons and protons and give them kinetic energy. This interaction works both ways; accelerated electrons produce electromagnetic waves.

Energy is not matter. It is not non-matter either, it is just a useful bookkeeping quantity. Energy is the measure of the work that is done or can be done in given conditions. In mathematical terms energy is force multiplied by the length of travel against the force.

Energy is conserved in closed systems. For instance, when we push a defunct car uphill, we work against the force of gravity, the weight of the car. The work done in the terms of energy is the car's mass times the height of the hill. If the car is then released and let to roll down the hill, the same bookkeeping amount of energy will be transformed into kinetic energy, which will then be released in an eventual collision. The end result will be heat energy; the random motion of atoms and molecules.

According to Einstein, energy and matter are related; the rest energy of matter is mass multiplied by the square of the speed of light. Einstein's famous equation $E = mc^2$ does not state that mass is energy, it only indicates their relationship.[1] Einstein's equation is experimentally verified.

Non-matter exists in modern physics. But this non-matter is not related to the philosophers' spiritual immaterialism and non-physical substances.

The philosophers' immaterial substances were supernatural; belonging to the realm of spooks, ghosts and spirits, and as such, not accessible to and explainable by known science. There is no empirical proof that supernatural substances and phenomena existed. On the other hand, the existence of matter should be self-evident. Matter is all around us, and our bodies are a part of the material world.

The denial of the immaterial effectively states that the brain is a material organ, and it does everything that it does by physical and chemical processes. No spiritual or immaterial component is involved, and consequently, there are no interactions between the material and immaterial to be explained.

---

[1] E = energy, m = mass, and c = speed of light.

According to materialism, matter is nature's sole substance, and everything that exists or happens in the world is based on matter and its interactions, also including radiation and force fields. This is seen to apply to mind and consciousness as well: Mind and consciousness are produced by the brain via electrochemical processes occurring in the brain's cells, neurons and their interconnections.

Indeed, there is ample empirical proof that brain's neural activity is connected with the processes of sensory perception, imagination, thinking, memorization and even the conscious and non-conscious states.

Materialism solves the problem of two substances and it also solves property dualism's emergence problem, but it does not inherently solve the problem of the apparently *immaterial qualitative appearance of mental contents* and the *problem of the observer self.* If these problems remain unsolved, the philosophical theory of materialism is not any better than the various theories of dualism and idealism.

3.5. The Philosophy of Dualism Debunked

Videre est credere — seeing is believing. But so is also the inability to see. Magicians' illusions are created by keeping things hidden.

Is it really impossible to explain the immaterial appearance of the mind by material processes? If it were impossible, would that prove that the mind is immaterial? Not really. The unexplainability might result from our ignorance. But on the other hand, a solid explanation of the immaterial appearance of the mind without resorting to any immaterial substances would solve the mind–body problem, and that would be a fatal blow to every kind of dualism. This is how it goes.

Everything is not what it seems to be, and impressions should not be taken at the face value. Magicians' illusions are created by keeping things hidden, so that the audience can be led to believe in something that does not exist in reality. The same applies to the mind–body dualism. The mind is not able to perceive the brain's neurons and neural networks as such because it does not have internal sensors to do that; the actual neural processes inside the brain remain hidden. However, the sensations produced by senses, thoughts, imaginations and other mental content are consciously observed, as our everyday experience shows. The impression of the immateriality of the perceived mental content arises from the mind's very inability to observe the material neural machinery. Due to this inability the machinery remains hidden.

This situation led to the early philosophical reasoning: The materiality of the brain cannot be denied, and the internal immaterial appearance of the conscious mental content cannot be denied either, as it is our personal impression and experience. Therefore, the immaterial mind has to be real, and of some separate immaterial substance. This conclusion led to the philosophical concept of substance dualism and the mind–body problem. Matter is matter, but what would be the immaterial substance? Possible immaterial substances, such as panpsyche and exotic consciousness fields have been proposed and sought, but never found due to the simple reason: They do not exist, and are not even required to explain something immaterial that does not exist either.

For thousands of years the philosophy of mind has tried to solve this mind–body problem, but even the wisest philosophers have not been able to find a credible solution. That should not be surprising, considering that they have tried to solve a problem that does not exist in the first place; philosophers have just fallen for the magician's curtain trick that creates illusions. Impressions may deceive, and the mere impression of immateriality does not prove logically and practically that something immaterial existed. Immateriality is a concept invented by humans, and the mind–body problem of dualism is an illusory problem. The early philosophers should have only noted that they are not able to observe the material mechanisms behind the mind. They did not realize this, and much ado about nothing ensued.

Things that are hidden from us are immaterial. That is a ridiculous idea. And a logically unsound one it is, too. Yet, this very idea is behind the mind–body problem. It is also behind the mystery of consciousness, and especially behind the hard problem of consciousness. And it will remain so as long as we try to explain the impression of apparent immateriality by the *existence* of something, instead of trying to explain the *origin* of the impression.

The hard problem of consciousness is related to the qualia, the apparently immaterial inner appearances of our sensory percepts. The apparent immateriality of qualia results from the mind's inability to perceive the material machine behind, and that is it. The remaining problem is the qualitative appearances of qualia; how are these generated? So far there has not been any acceptable material explanation of qualia. This is unfortunate, as all the mysteries of mind and consciousness would evaporate as soon as the neural basis of the immaterial qualitative appearances of qualia were explained. This explanation is provided in the Chapter 8 of this book.

But who observes qualia? That is the *observer problem*. How does the mind observe its contents and create the experience of self?

Chapter 4

The Observer Problem

4.1. Perception and Mind

Our senses provide us information about the external world. For instance, we hear sounds with our ears that transmit neural signals to the brain and the experience of heard sounds follows. We see the world around us with our eyes that operate like small cameras and project the image of the seen world on the receptor neurons of the retina at the back of the eyeballs. Millions of sensory signals are transmitted from the eyes to the brain, and the direct appearance of the seen world is generated. All our senses produce their sensations in similar ways producing direct sensory experiences and appearances with the apparently immaterial forms of qualia.

This leads to the unavoidable question: Appearances to whom? Who or what is observing these appearances? The obvious answer is: The observer is the self. But what is that kind of an observer self, and where would it reside inside the brain? And how would the self do the observing?

Mind–body dualism explains that the observer is the immaterial soul, and this solves the problem. Materialism maintains that there is no immaterial observer — therefore the observer problem arises again. This time the problem is not only a philosophical one, it is also a technical one relating to the research of machine consciousness.

Technically, how would the mind and self observe the inner appearances? The observer problem is at the root of many problems of self and consciousness in the philosophy of mind. This problem is solved in the following.

4.2. The Problem of the Observer and the Experiencer

An early attempt to explain the workings of the senses and the brain was the homunculus metaphor. It was thought that in the brain there was something like a theater stage or a movie screen, on which the eyes projected their colorful imagery to be observed by the conscious mind. The observer mind was seen to have the form of a small person, a homunculus. The homunculus would be the actual self with its own mind. This metaphor is known as the Cartesian theater [Dennett, 1991], and as a metaphor it would seem to work well enough.

Well enough is seldom well enough. The Cartesian theater has two problems; the screen and the homunculus observer. Where is the screen in the brain, and what is the homunculus, and how does it see what is on the screen?

If we are to accept the Cartesian theater model as the explanation for the observer problem, then we have to present how the homunculus itself works. It is simple; inside the homunculus there is another screen, a smaller one, and also a smaller homunculus, who is observing what is on the smaller screen. The problem is solved, but then there are the smaller screen and homunculus calling for similar explanation. And so forth, until the final observing homunculus is reached. And that, apparently, would be so far away in the infinity that it would not need an explanation, maybe, see Fig. 4.1.

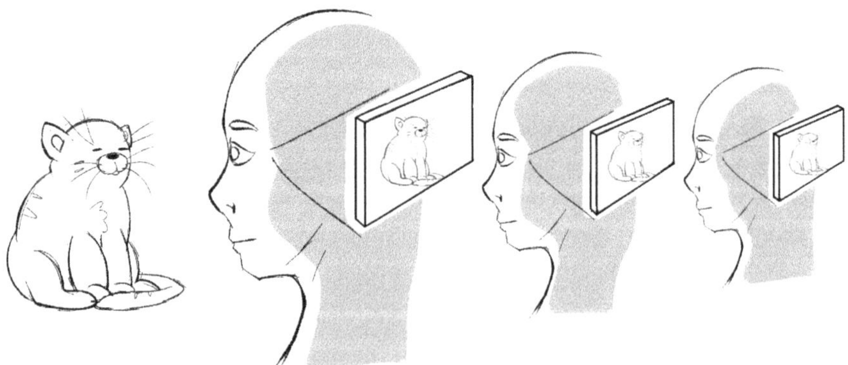

Fig. 4.1. Cartesian theater and the chain of smaller and smaller observing homunculi inside the head.

Figure 4.1 depicts the succession of Cartesian theaters and homunculus observers. Do the successive observations of the observed solve the observation problem? The obvious problem here is the second observer; how does this observer observe, does it also have senses that produce sensations to be observed by a further observer … and so on? Or, would there be a final observer, who observes directly without the need for further "theater stages". If that kind of observer were possible, then what point would there be to have any previous theater stages or "display screens", for that matter? It would be much simpler, if the world itself were the "theater stage" for the senses to perceive and observe.

The homunculus has been rejected as a credible solution to the observer problem, which thus has remained unsolved. Nevertheless, Cartesian theater stage models are still seen in modernized forms in some models of consciousness, and also some kind of regression in the brain's neural networks are sought for as an explanation for the "emergence" of qualia, the inner appearances of sensory percepts.

The observer problem may appear to be so difficult that it may well be without a natural or material answer at all. But sometimes, difficult questions are difficult because they are wrong questions. The internal observer question may be such a question. If so, it does not even have to have an answer — and it is explained here why it is so.

The observer problem remains as a difficult question to be answered to, as long as we are searching for a definite observer entity, the one watching and observing what the senses deliver. No Cartesian theater observer entity has been found because apparently, there is no such entity. Thus, the search for an inside observer seems to be futile and should be abandoned. But then, how could the mind perceive anything without being an observer? But musings like these are of no help, and the homunculus fallacy is still haunting there.

However, there is a solution, and it is an obvious and simple one that does not require any internal observer entity.

The key here is the difference between the insider view and the outsider view. When one is observing what is outside, one is an outsider. But in respect to what is inside, one is an insider. An outsider is an observer, while an insider is not; an insider is an experiencer. The point is: What is outside, can be observed. What is inside, can be experienced.

For example, we can painlessly observe that others are in pain, just because we are outsiders. But when we are in pain, we experience it, and this insider experience does not call for any further act of observation in order to become the feel of pain.

The above applies to all modalities of sensory perception. Sensory perception facilitates the observation of what is outside by producing sensory percepts, which are neural excitations of the sensory perception neurons inside the brain. In respect to these we are no longer outsiders, we are insiders, and these neural excitations are not observed; they are experienced.

All our percepts of the external word like the sensations of colors, sounds, tastes, etc. are inner experiences. As such they are directly parts of the experienced mental content; there is no need for any additional observation process or an observer.

Neither do we observe our thoughts as such; they are not something that we would have to somehow inspect via some observer act in order to become aware of them. Thoughts are inside, they are experienced and are already a part of the conscious mental content (Introspection is related to the meanings of thoughts and emotions; that is not the issue here).

The Cartesian theater model does not utilize the insider-outsider concept. Instead of this, sensory percepts, thoughts and imaginations are presented on the theater stage to be observed by an observer entity. This does not remove the need for further observing, and then for one more, and so on ad nauseam. The Cartesian theater model is truly faulty.

What is inside, is experienced. What is outside, is observed. The understanding of the insider experiencer — outsider observer role division is a cornerstone in the understanding of the issues of the mind, consciousness and especially in the explanation of qualia, the qualitative inner appearance of the neural signal patterns and excitations of percepts. See the following chapters for details.

Chapter 5

Qualia, the Evasive Essence
of Perception

5.1. Perception and the Concept of Qualia

The evasive essence of perception that pivots everything.

Our sensory perception produces a variety of direct experiences. What we see, we take to be as what they are. What we hear, we take as the sounds that we hear. What we taste and smell, we take as the tastes and smells of the sensed substances. The experience of pain is the pain, the experience of pleasure is the pleasure.

Our sensations or percepts have different qualities that are related to the sensed entities. The taste of coffee, the sweetness of sugar, the smell of a rose, the blueness of the sky — they all are familiar examples of the experienced qualities of what is sensed.

All this happens so directly and effortlessly that we do not pay any attention to it, and we are not even aware of the neural machinery that executes whatever has to be executed in order to produce the sensory experiences with their qualities. The neural material machinery is there, but remains hidden; we are not aware of it as it is. Instead of any awareness of the neural machinery, we are very much aware and conscious of the perceived information. On the basic level, the world outside us seems to be self-evident and self-explanatory, with its large variety of perceived qualities.

In order to bring these sensed qualities into a proper philosophical focus, an American philosopher Clarence Irving Lewis (1883–1964) introduced the term qualia (singular quale, from Latin quality) in 1929 in the context of sensory perception and the philosophy of mind. Lewis proposed that qualia would be a name for the recognizable qualitative characters of the sensory information, see e.g. [Goodman, 1977].

Lewis' term qualia was widely adopted. As a result, many books on the philosophy of mind treat this subject, and a vast number of papers have been written and published about qualia. Now there is no learned talking about the philosophical issues of mind without any reference to qualia. But there is a problem; instead of clarification, the concept of qualia has caused a great confusion. What, if anything are qualia? Are they properties of the sensed items or are they depictions of the properties of the sensed items? For instance, is the saltness of salt a property of salt or only our sensory impression?

"Qualia" is an ill-defined concept. This is a fact, demonstrated by the many, many, more or less contradictive definitions for qualia given in literature and in the web. Apparently, many individual philosophers have their own concepts and definitions, which are not explicit to others.

Are qualia immaterial? Australian philosopher Frank Jackson presented that qualia are perceptual experiences that cannot be produced by physical information only; thus qualia have to be immaterial [Jackson, 1982]. On the other hand, American philosopher Daniel Dennett argued that qualia do not even exist [Dennett, 1991] and became famous via this. It has also been proposed that qualia and consciousness are just illusions caused by the neuronal activity of the brain. But then, this should not go without asking that to whom would qualia be illusions, and what exactly would these illusions be, and what would be the difference between illusions and inner appearances.

These and all the other views about qualia make serious discussions challenging, as it is difficult to know what people are actually talking about. The words may be recognized, but what has been meant may remain vague. This has resulted in never-ending disputes about the meaning and essence of qualia.

Examples of qualia (or anything) do not constitute explanations, and definitions via examples leave much to be desired. Therefore, a better definition for qualia would be required, one that would clarify the concept and allow rational analysis behind the mystery of qualia.

5.2. Qualia Defined

Good definitions are indispensable steps toward good explanations, but it should be noted that definitions in themselves do not constitute explanations.

Properly understood and defined qualia have a pivotal role in the solving of the mind–body problem, in the explanation of consciousness and in the solving of the symbol grounding problem.

The following example from visual color perception may illustrate the case of qualia. We can see that the sky is blue, but why? The answer is not as simple as might be assumed. The sky is blue because air molecules scatter blue wavelengths of the sunlight, and consequently blue light appears to originate from everywhere in the sky. This is a good, short and simple explanation that presents a physical fact. But, it does not actually explain why the sky is blue.

Photons do not have colors; they only have different radiation energies. Our eyes do not see colors either, as colors do not exist. Our eyes respond only to the photon energies, and transmit their neural responses to the visual cortex of the brain. But we do not experience these neural responses as what they actually are; electrochemical neural activities. We see them as colors. These colors are the subjective way in which these neural responses appear to us internally. Moreover, colors are not the qualities of light or seen objects, they are just subjective sensations that are evoked by photons entering our eyes. Colors are one part of visual qualia.

All this is related to the essence and problem of all kinds of qualia. All our senses react to their stimuli and produce their specific neural responses, which are transmitted to the brain. However, these neural responses do not appear internally as what might be expected, namely as neural activities in the brain. They are not experienced as neural activities, but instead, they are taken as qualities of the sensed; they are qualia.

Qualia are not representations for the mind to inspect and decode. They are taken directly as the sensed thing; red is red, cold is cold, pain is pain, no interpretations or explanations are needed. Qualia are self-explanatory.

All our mental content appears to us as sensory percepts of the external world and as virtual sensory percepts of the mental content, like the silently heard inner speech, virtually seen imagery, emotional states, pain and pleasure. They all have their different qualities, and manifest themselves as they are, in a self-explanatory way. Therefore, instead of trying to define qualia via lists of qualities and miscellaneous properties, it would be more constructive to realize what they really are. Regarding this, qualia can be defined as follows:

> *Qualia are the way in which all our conscious mental content appears to us. There is no consciously perceived mental content without qualia.*

This definition links qualia with the phenomenon of consciousness, as it presents that all consciously perceived mental content has the form and appearance of immaterial qualia. Therefore, the solving of the fundamental problems of qualia, namely their impression of their immateriality and apparent qualitative appearances, would also solve the fundamental problem of consciousness and the immaterial mind — material body problem, see Chapter 18. Also, as qualia are self-explanatory, they will also facilitate the solving of the symbol grounding problem, see Chapter 13.

Definitions are not explanations, and this applies also to the above definition of qualia. The true explanation of qualia would explain how qualia arise in the neural workings of the brain and in the mind, and it would also explain their qualitative and immaterial appearance.

This kind of an explanation of qualia is given in the Chapter 8. This explanation is based on neural activities, and therefore it calls for some basic knowledge about sensory perception, the brain and neurons. The required issues are treated shortly in the following two chapters.

The neural explanation of qualia will be a central factor in the author's explanation of consciousness, and this situation will lead to some unexpected conclusions about the existence of consciousness.

Chapter 6

The Brain and Neurons

6.1. The Brain

No brain, no gain, no mind.

In the external appearance of the brain there is nothing that would hint towards the mental life inside. Brain is matter, how could it possibly host an immaterial mind or even generate the impression of a mind? How does the brain generate vivid qualia, thoughts and inner speech?

Nowadays it is possible to peek inside living brains by various imaging technologies, like the Magnetic Resonance Imaging (MRI). The overall structure with some details of the brain is revealed, but not the mental workings. A typical MRI image of a brain is depicted in Fig. 6.1.

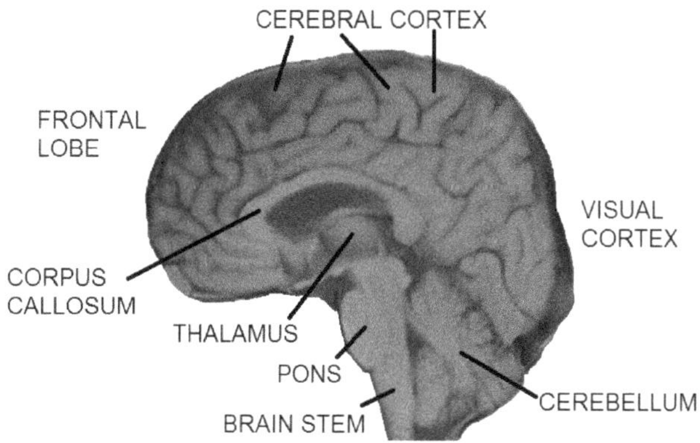

Fig. 6.1. A MRI image of the author's brain (by courtesy of the owner).

There are also other instruments that are able to detect electro-magnetic activity patterns, which may be correlated with percepts and thoughts.

The brain is the main component of the so-called central nervous system (CNS). Other parts of this system are the spinal cord, and the two-way neural wiring between the brain and sensory organs and muscles. Via these connections the brain acquires information about the environment and the body and controls the body.

To understand the workings of the brain one has to understand the general workings of the living cell and the brain cells; neurons.

6.2. Living Cells

A biological cell is the building block of all living organisms. A living cell needs energy, and acquires that via metabolism; the intake of energy producing molecules and the ejection of waste molecules. The living cell is also able to heal and repair itself. Basic living cells multiply by division.

A typical biological cell consists of the cell membrane that shields the cell's innards. The main items inside the cell are the cytoplasm, the nucleus, ribosomes and mitochondria, see Fig. 6.2.

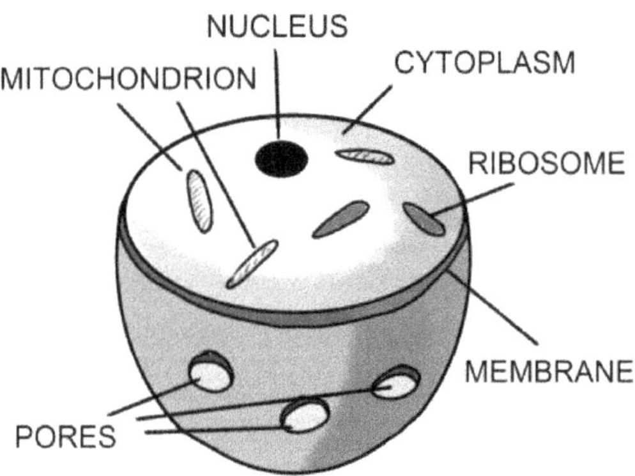

Fig. 6.2. The outline of a biological cell (cut).

The cell membrane forms a shielding wall around the innards of the cell. The cell membrane is not completely closed; there are extremely tiny openings, so-called pores, in the cell wall. The pores allow selected molecules and ions to pass in and out of the cell. Pores facilitate metabolism.

Cytoplasm is a semi-fluid liquid that contains proteins, nucleic acids and molecules produced internally or taken from blood circulation.

The nucleus is a small, specialized structure that contains DNA. DNA (deoxyribonucleic acid) is a long helical double-stranded structure. DNA contains hereditary material that allows the reproduction of the cells and the whole being. Short sections of DNA constitute genes and other building plans for the whole organism.

Ribosomes are platforms for protein synthesis. Mammalian cells may have up to ten million ribosomes.

There are around 1000 to 2000 mitochondrion cells (Plural: mitochondria) in an animal cell. Mitochondria are kinds of parasites, as they are cells within the cell and they have their own DNA. However, mitochondria are useful, as they live in a kind of symbiosis with the host cell by transforming nutrient molecules into forms that the cell can use more effectively.

Cells execute various tasks in living organisms. They synthesize proteins of various kinds, according to the activated genes. These proteins are used as building materials for the body and also for other purposes. Specialized cells form individual units such as internal organs and also the brain. Cells multiply by making copies of themselves by division.

A living cell has many things going on inside at every moment.

6.3. The Neuron

Neurons are brain cells and are basic biological components of the brain [Nichols *et al.*, 1992, Thompson, 1993]. They are excitable cells that fire electric impulse trains (neural signals) when sufficiently excited. Their overall function in the brain is to receive information, connect it with other information, store connections and memories and do whatever is needed for cognition.

Figure 6.3 depicts a simplified two-dimensional sketch of a neuron, its axon with the axon terminal bulb, dendrites and some synapses. Normally there would be a large number of synapses.

Neurons receive and transmit neural signals. These signals consist of electric impulse trains, where the individual impulses have constant amplitude voltage (ca. 30 mV). The impulse repetition rate varies and in this way reflects the excitation state of the neuron. Non-excited neurons transmit random impulses every now and then.

Neural signals arrive and leave the neuron via axons, which are very thin hollow nerve fibers. The input signals are connected to the neuron

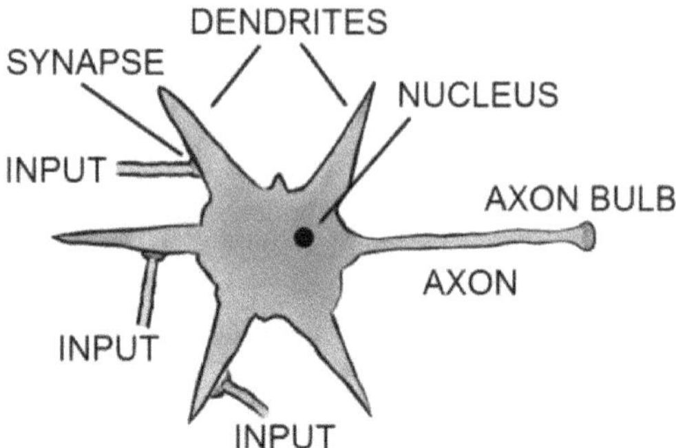

Fig. 6.3.   A simplified sketch of a neuron with axon, dendrites and synapses.

via so-called synapses, which consist of the axon terminal bulb and the connection area on the neuron body. There is a gap between these; a kind of threshold.

Simple models of biological neural networks present that the logical operation of the neuron would be based on the electric behavior of the neuron; neurons would receive trains of electric impulses, and if thresholds were exceeded, they would transmit their own response in the form of electric impulse trains via their output axons. Simple logical functions can be executed in this way.

It is understood that networks of neurons learn and do whatever has to be done for cognition.

6.4.   Synaptic Learning

It is generally hypothesized that in the brain the memory traces are created by the modification of synaptic transmission strengths of individual neurons. Long-term and short-term potentiation and synaptic plasticity are often used as the names for the modification of synaptic transmission strengths.

The actual synaptic mechanism is not yet completely understood, but it may be something like this. An agitated synapse transmits neurotransmitters to the dendritic receptors of the receiving neuron, and in this way communicates its message, namely the presence of the agitating neural signal. The synaptic transmission may be initially weak, but may get stronger via repetition, see Fig. 6.4.

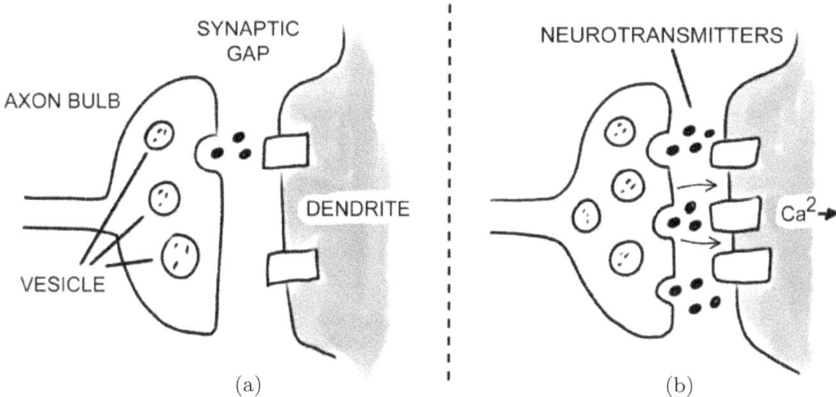

Fig. 6.4. Repetition modifies synaptic transmission strength. (a) Initial state showing weak synaptic transmission. (b) Strengthened synaptic transmission following repeated stimulation.

The synaptic transmission has two preconditions that are the presence of neurotransmitters in the axon bulb vesicles of the transmitting neuron and the presence of dendritic receptors in the receiving neuron. Without neurotransmitters there is nothing to be released and transmitted, and without dendritic receptors nothing can be received. In both cases there will be no synaptic transmission. During learning the amount of neurotransmitters and dendritic receptors may increase (It should be noted that the astrocyte cells may also have their role here. This is discussed in Chapter 12).

6.5. Neurons as Pattern Recognizers

Pattern recognition is a simple task involving learning and memory, and this process can be explained by the synaptic transmission strengths of individual neurons. For instance, when we repeatedly see a given object, we will become familiar with it. Afterwards we will recognize it, when we see it again. No names are necessary for this.

According to a simplified neuron model, neurons execute simple pattern recognition as follows. In the brain sensorily observed patterns appear as patterns of neural signals. These signals will be the synaptic inputs for a pattern learning neuron. Initially the synapses would not pass the signals, but after some repetitions, the synaptic transmission strengths would increase and the signals would pass.

The neuron body operates as a summing device with threshold. The excitations from the individual synapses are summed together. When the sum exceeds a set threshold, the neuron fires the output signal.

This would happen, whenever the same neural signal pattern is encountered; the pattern has been learned, and the neuron's output signal would indicate the detection of the learned signal pattern. It should be noted that the simple sum and threshold operation alone does not guarantee correct recognition. The effect of missing and improper synaptic inputs must be taken in account, see Haikonen [2019].

Neurons that respond to only one definite stimulus pattern are sometimes called "grandmother" or "Jennifer Aniston" neurons, as they would output a signal train whenever they detect their object. These kinds of neurons have been actually found in the brain, and single signal encoding may have an important role. Nevertheless, there is a lot of redundancy in the brain, thus there may also be many neurons that execute the same tasks.

Groups of parallel neurons can be used to recognize different neural signal patterns.

Chapter 7

Sensing and Neural Codes

7.1. Cells and Neurons Sensing the World

All biological cells interact with their environment, and are affected by external conditions, such as light and radiation, temperature, chemicals, and mechanical impacts including pressure, acceleration and vibrations. All these have their effects on the internal processes of the cell.

There are also specialized biological cells that are excitable by specific stimuli, and are able to transmit their response to other neurons. These cells are known as receptor cells. Receptor cells facilitate sensing, and they respond to various external and internal stimuli. Sensing is reacting to stimuli; no reaction, no response, no sensation.

Receptor cells are sometimes presented as transducers, which transform their specific stimuli like touch, light, sound, into electric and chemical neural signals. This is seen necessary, as obviously the brain can only utilize information carried by neural signals and chemical messages.

Actually, this is not accurate. Consider the mercury thermometer that is supposed to measure and indicate the room temperature. The mercury thermometer has a thin glass tube that contains a small amount of liquid mercury. The mercury inside expands and contracts according to *its own temperature*, and the temperature can be read from an adjoined scale. Nothing is actually transduced here, as what is indicated, is the quantity of the expansion of the mercury. All measuring instruments operate in the same way, and the produced value actually indicates the state of the sensing element, not anything else. Usually that does not make any difference.

Biological receptor cells are not different in this respect. They react to the received stimuli, become excited and transmit their neural signal response. Our sensory impressions are based on the receptor cells' excitation states, nothing more. Mind operates with these experiences of excitations.

The transmitted signal does not convey any information of this origin. *The what* and *the where* have to be determined by other means, as is described later on here.

Humans and animals have a variety of receptor cells that are sensitive to different stimuli. These receptors can sense and deliver mechanical, auditory, visual, temperature and chemical information. Systems of these receptors form the sensory nervous system.

There are two kinds of sensing; contact sensing and sensing from a distance. Sound sensing and visual sensing are examples of latter.

Touch sensing is contact sensing, and it is performed by mechanoreceptors, which are located under the skin. When skin is touched, these receptors respond to the pressure and send a signal to their sensory neurons. There are many kinds of mechanoreceptors, one of them is hair. Hair itself does not sense anything, though. The sensing is done by the nerve fibers around the hair bulb at the root of the hair. When hair is bent, the signal is sent. That is easy to confirm; just gently move your hand over your hair.

Proprioception is the name for the sense of the spatial position and movement of body parts in relation to each other. For example, it allows the perception of the position of hands without the help of vision. One can close eyes and still know where the hands are. This function is enabled by proprioceptor cells that sense the tension of skeletal muscles. Proprioception is necessary for the control of movements.

Haptic perception is the perception of the surfaces and forms by explorative touching. It utilizes both touch sensing and proprioception. Haptic perception generates inner three dimensional "image" of the sensed object by sustaining and combining the instantaneous sensations of touch and position. This "image" can be memorized.

Nociception is the name for the sensing of temperature and cell damage. It can lead to the experience of pain.

Sound sensing is sensing from a distance. Sound propagates in the form of sound waves, which are miniscule air pressure vibrations. Air itself is not detected and air molecules are not transmitted, only the vibrations are. The vibrations are not matter, they are only propagating density variations of the air. These vibrations are detected by microscopic sensory hairs in the hearing organ, the cochlea, which resides in the inner ear. The excitation of cochlea leads to the experience of sounds. As no matter is received from the sound source, also sound sensing can be seen as immaterial.

Visual sensing is also sensing from a distance. This does not involve any material contact. No matter is received from the sensed source; only light is sensed. Light consists of photons, ripples in the electromagnetic field. Photons are non-matter and consequently are not observed as such, it is their energies that have the effect on the material eye. In the eye's photoreceptor cells photon energies produce various chemical compounds, agitating thus the receiving neuron cells [Goldstein, 2002]. As no matter is received from the source of the observed object, visual sensing can be seen as immaterial.

7.2.   The Quest for the Neural Signal Code

In sensory perception, neurons in the brain's sensory perception areas receive neural signals from the various sensory receptor cells, and this in turn leads to the various meaningful sensory experiences in the mind, see Fig. 7.1.

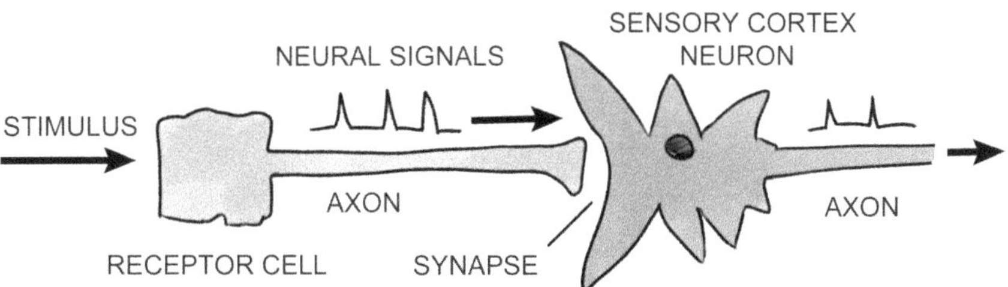

Fig. 7.1.   Sensory signal transmission from a stimulated receptor cell to the receiving sensory cortex neuron in the brain.

Figure 7.1 depicts the basic operation of the signal transmission from a sensory receptor cell to the sensory neuron in the corresponding sensory cortex area. Here, an outside stimulus excites the receptor cell causing it to fire an output impulse train, which travels along the axon to the connecting synapse of the sensory neuron. Obviously, in this way information is transmitted from the receptor cell to the sensory neuron. But how is this information carried by the neural impulse train?

At first sight it would appear that the transmitted information from the sensory receptor cell would be somehow modulated or encoded in the electric neural signal; there would be a neural code [Johnson, 2000]. This idea has led to various neural code theories. According to these theories the electric neural signal carries coded information, which is then decoded in the receiving neural assemblies. Also, the deciphering the code might allow the external access to the meanings carried by the neural signals.

It is not immediately clear what the code would be in the neural signal. Neural signal impulses have constant amplitude, and therefore cannot carry any information in that way. Therefore, the code must be carried by other means, like the number of impulses in each burst of impulses, the repetition rate of individual impulses, correlations between impulses, and so on. These variables and their patterns would represent whatever is transmitted, and would be decoded in the brain, see e.g. [Li and Tsien, 2017].

Neural code theories present how information may be transported from a neuron to another in a coded way, but unsolved questions remain: How does the encoding and decoding take place? That is a problem.

In the neural signal impulse train, the repetition rate of the impulses is obviously related to the intensity of the stimulus, and the repetition rate of the bursts of the neural impulses would follow the rhythm of the stimulus, but where is the actual code?

Lot of research has been done along these lines, but no unambiguous neural code has been found. There may be a good reason for this: The transmitted electric impulse train does not actually carry any encoded information; it is not the transmitted message at all.

Nevertheless, the trains of impulses are real and can be detected. So, where is the message, if it is not in the electric impulse train? — That will be revealed by a more detailed inspection of the operation of the synapse.

Typically, a sensory system consists of a receptor cell assembly and a sensory neuron assembly inside the brain. Receptor cells are sensitive to their specific stimuli, and they will react when they are stimulated. As a result, they will transmit signals via their axons to their sensory neurons at their corresponding brain areas. But the electric impulse train is not the only signal; the signal that affects the receiving neuron is chemical.

It has been found out that the axon fiber terminal bulb contains small vesicles that contain so-called neurotransmitter molecules. Basically, there are small and larger neurotransmitter molecules. Small ones are synthesized and stored in the terminal bulb and can be released quickly.

Larger ones, so-called neuropeptides, are synthesized inside the cell body, and are conveyed through the axon to the terminal bulb.

When a suitably strong electric impulse train arrives at the bulb, minute amounts of neurotransmitters are released, and they will cross the receiving neuron cell membrane. Thus, the actual signal is a chemical one, carried by the released neurotransmitters. In this way the electrochemical excitation state of the transmitting neuron can affect the receiving neuron. There is no encoding, there is no code, there is no decoding. There is no enigma, see Fig. 7.2.

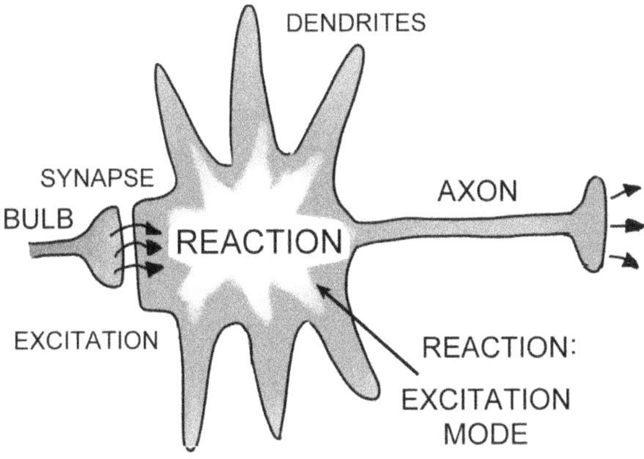

Fig. 7.2. Synaptic release of neurotransmitter molecules leads to the receiving neuron reaction and excitation mode.

Traditionally, it was assumed that each transmitting axon bulb would be able to deliver only one kind of neurotransmitter. Recent research has revealed that this is not the case, see e.g. [Vaaga *et al.*, 2014]. The axon bulb may contain two or more different neurotransmitter vesicles, or one vesicle may contain several different neurotransmitters. The released neurotransmitters may be selected by the intensity (or perhaps the envelope form) of the electric impulse train and also by chemical messages. In this way the receiving sensory neuron may react to both intensity and quality of the received stimulus, instead of mere intensity, and have a larger repertoire of *neural excitation modes*.

Inside the cell, neurotransmitters will affect the internal condition of the neuron; different excitation states or modes will follow depending on the received neurotransmitters and their amounts.

In summary: The connections between sensory cells and sensory perception neurons are not about the transmission of any messages, they are about stimulation. There is no neural code, no encoding, no decoding. The receiving neuron decodes nothing; it has no idea what the stimuli are about, and it only responds to the received stimuli by its own reaction.

There is no neural signal code and it is not needed for anything. This simple conclusion makes surprising sense in the context of sensory perception, neural associative information processing and the perception of mental content.

Especially, in the context of associative processing, the transmitted associatively evoking neural signal patterns do not carry any meaning, they only indicate the presence of the cause of the signal pattern. It will be seen that apart from the early sensory perception neurons, only the presence of signal patterns matters, not what the signal pattern is about. It is only about making and activating connections, while the actual meanings and qualia reside only at the early sensory perception neurons, see Chapter 14.

Chapter 8

The Material Explanation of Qualia

8.1. The Neural Basis of the Inner Appearances of Qualia

The solving of the mystery of the immaterial inner appearances of qualia also solves the hard problem of consciousness. The mind–body problem is also put to rest, as the verified existence of material processes behind the immaterial appearances negates the requirement for any immaterial substances.

In this book, qualia are defined as the way in which all our consciously perceived mental content appears to us internally. This applies to both sensorily produced and internally produced contents of the mind.

However, this definition (and any other definition of qualia for that matter) would be worthless without any explanation of the processes behind the qualitative and apparently immaterial appearance of qualia.

Introspection shows that the mind's internally produced contents, like the inner speech and imaginations, have the form of vague virtual sensory percepts with qualia. We "hear" our thoughts and "see" our imaginations. Inner speech is rather similar to heard speech, and imaginations are likewise similar to actually seen imagery. Why this is so is explained later in this book. Therefore, the explanation of the origins of sensory qualia should suffice.

The brain is a biological neural network, and senses are biological organs, which deliver their outputs to the brain in the form of neural signals. Therefore, the material explanation of qualia should be found in the workings of neurons.

Basically, all sensing is based on receptor cells (sensors). When these cells are stimulated, they will respond and produce neural signal outputs, which are then transmitted to the sensory cortex areas in the brain.

There are many varieties of receptor cells, like those responding to light in the eyes, those responding to sound in the ears, those responding to touch, and so on. When stimulated, all these respond in the forms of their own kinds, and corresponding qualia follow.

8.2. Amodal Qualia

If we were only able to know the sensory neuron excitation mode and the corresponding quale, we would be a little step nearer to the solving of the qualia problem.

The concept of amodal qualia helps here to begin with. In certain cases, the observation is the same as the observed. In that case, the quale is actually similar to the observed entity, and these kinds of qualia are also the same across various sensory modalities. These qualia are called amodal qualia[1] [Haikonen, 2019]. Amodal qualia include felt and seen shapes and also felt and heard rhythms.

A simple experiment in the area of touch sensing may illustrate the usefulness of this idea. Tap your table with your forefinger shortly. You will experience a fleeting sensation, perhaps the simplest quale.

When you tap your table, the mechano-receptor cell in the tip of your finger sends a transient impulse signal to your brain. This impulse causes a corresponding transient internal effect in the receiving sensory neuron. This effect is the "tap-like" excitation of the neuron.

We know the chain from the stimulus tap to the resulting neuron excitation state. We also know the resulting quale because that is what we experience. Here, the beauty of this is: The tap-like quale is amodal. Thus, if there is nothing else, then the sensory neuron's transient excitation state must be the quale with the feel and all. The neuron itself is the experiencer.

---

[1] Amodal qualia are not to be confused with the so-called amodal perception, e.g. the virtual perception of a partly hidden object as a whole.

It experiences the "up–down jump" of the tap and the series of "up–downs" of repeated taps.

The "tap" quale, as well as all other qualia, can be sustained and remembered for a while with various mechanisms. This is necessary for the detection of change.

8.3. Auditory Qualia

Next, a more complex example of qualia is presented, in this case in the modality of auditory hearing. The process of hearing is depicted in Fig. 8.1.

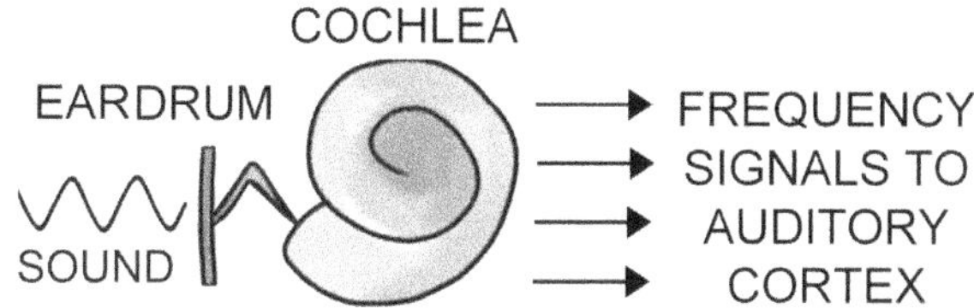

Fig. 8.1. The path from heard sound to sensory neuron reaction.

Figure 8.1 depicts the signal path from the sound stimulus to the auditory cortex neurons. Sound is actually minute vibrations of air pressure. In the ear, the eardrum transforms these vibrations into mechanical vibrations, which are coupled to the inner ear organ called cochlea. This is a coiled hollow cone, which is filled with fluid. Received sound vibrations propagate in this fluid. The cone also contains rows of hair cells (the organ of Corti), which vibrate along the vibrations in the fluid. These hair cells are able to convert the mechanical vibrations of their hair into neural signals. Each hair cell resonates only with one frequency and will deliver its neural response signal when this frequency is present. In effect, the cochlea performs audio frequency spectrum analysis, giving the spectrum of each heard sound. Young persons with normal hearing may be able to hear sound frequencies from ca. 20 kHz down to 20 Hz or even lower.

The sounds that we hear normally consist of sums of many frequencies, they have a spectrum. Without any spectrum analysis, the identification of different sounds and the separation of individual sounds from other simultaneous sounds would be very difficult. The cochlea solves this problem.

Single frequency signals have the form of the sine wave, see Fig. 8.2.

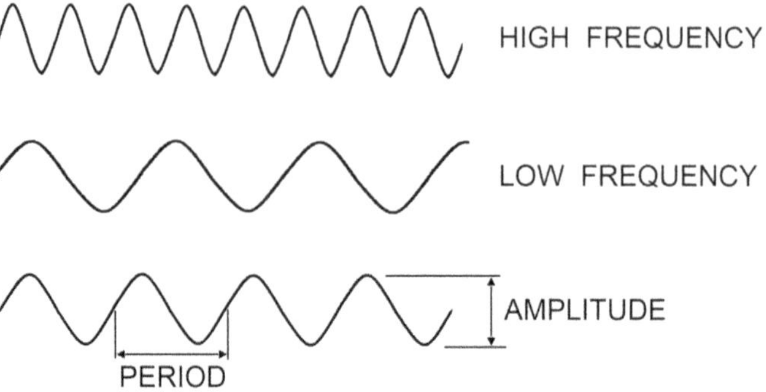

Fig. 8.2. Sine waves with different frequencies.

Sine waves are the simplest periodical waveforms, consisting of only one frequency. The frequency of the sine wave is determined by the duration of the period (cycle) of the sine wave, consisting of one hill and one valley. All sounds are composed of sums of sine waves.

An experiment may illustrate the connection between the heard sine wave sound and the resulting auditory qualia. This experiment utilizes an audio signal generator capable of producing clean sine wave audio frequencies from practically zero (DC) to 20 kHz or more. The person doing this experiment shall listen to the audio signal with high-quality headphones with the ability to reproduce low-frequency signals.

During this experiment, it will be noted that the sensation of the lowest audio frequencies appears as the train of individual periods of the signal. This is the rhythmical repetition of hills and valleys, and the sound experience is like hump–hump–hump.

When the audio frequency is increased, it will be observed that the individual periods of the signal fuse together into a continuing whistle, while in reality, no such thing happens in the signal generator; the signal conserves its sine form and amplitude up to the maximum frequency available. The experienced fusing of the periods results from the slowness of the neural process; it is not fast enough to be able to resolve and report the individual periods of higher frequencies. Higher and higher frequencies will be heard as higher and higher-pitched whistles. These are the perceived qualia of individual tones.

It should be easy to realize what the sensory neuron at the auditory cortex is doing; its excitation state follows the periodic injections of the neurotransmitter. This is a physical phenomenon, which is behind the sub-jective hearing experience, the quale. This situation is not much different from the sensing of repeated mechanical taps; the experienced qualia are rather similar even though from different sensory modalities.

8.4. Visual Qualia

Cover your eyes in total darkness. What do you see? Random excitations of your photoreceptor cells.

Visual qualia are somewhat different. The simplest example of visual sens-ing is the setup of one light sensor cell and one sensory neuron. The light sensor delivers an output signal with a strength that is relative to the inten-sity of the sensed light. In this way, the intensity of the external illumina-tion can be detected. That is not much, but it would be valuable information for small critters, which live and hunt in darkness and try to avoid lighted areas, where the predators are.

Here, the sensed light intensity would stimulate the sensory neuron, and the sensory neuron would react to this by raising or lowering its excitation level. The internal "experience" of this excitation would be the quale.

Let's assume then that the light sensor had a very narrow angle of vision, which would allow a definite gaze direction. If the critter were able to point this light sensor toward different directions and at the same time observe the gaze direction, then the critter could map the light intensity at each direction. The critter could even form an illumination map of the environment if it were able to associate the gaze direction with the sensed light intensity and remember these for a while.

All this could be effected better if instead of one light sensor, there were a matrix of light sensor cells and sensory neurons, and some optics that would project an image of the environment on this matrix. The end result would be a pattern of sensory neuron excitations, "a grayscale image". This image would allow the determination of what is where. The patterns would be the conscious experience.

Color sensations could be effected (and are effected in the eye) by the use of three kinds of light sensor cells or receptors, one for red, green and blue visual wavelengths. In this case, the sensory neurons for each color would produce different modes of excitations and, consequently, different qualia.

The fundamental principles of sensory perception are quite simple, but seeing may appear as an awesome and mysterious process. This impression is created by the very high resolution of the eye, especially at the focus area in the middle of the retina, the fovea.

8.5.   Qualia as Experiences of Changes

A movie where nothing happens is not much of an experience.

Perception is very much about the detection of spatial and temporal changes. We see things against static backgrounds, and we hear sounds against silence. Odors and flavors are noted as they differ from the ordinary. We do not actively notice the state of our body as long as no pain is felt. Patterns are noticed as they differ from their background. Look around, and you will note that the seen visual details stand out because they differ from their surroundings.

Changes provide distinction. Nothing will be perceived if nothing changes. However, it could be argued that this cannot be so. For example, one could look at a photograph and see it effortlessly, even though definitely nothing changes in the photograph. Surely then no change is required. Nothing changes temporally in the photograph; that is true, as the changes in the image are spatial and fixed, not temporal. But our gaze direction changes temporally as we glance over the details of the photograph. This visual scanning causes the details of the photograph to be observed sequentially in time, thus the visual excitation on the retina in the eye will change temporally. There will be changes.

Sounds are temporal changes in the air pressure and are detected as temporal changes of neural excitations. Likewise, our touch sensations involve temporal change, as they change when we touch and feel objects.

The relevant point here is that our senses are sensitive to the changes in the stimuli. The changes are detected, and neural responses follow. In static conditions, the sensory receptors are in a rest condition, and their inner electrochemical processes are balanced. A received stimulus causes imbalance inside the receptor cell, and this imbalance will generate changes in the inner electrochemical excitation modes of the sensory perception neurons. These electrochemical excitation modes have their own qualities, which follow from the spatial and temporal properties of what is sensed. The qualities of the excitation modes depend on what is sensed and also on the different sensory modalities and their corresponding receptor properties. A certain amount of amodality may be involved in this.

Qualia are subjective qualities of what is sensed. The neural excitation modes are also related to the qualities of what is sensed, and therefore, it may be asked: Are the inner experiences of qualia related to the excitation modes of the sensory perception neurons, and could qualia be somehow equated with the excitation states? This question is addressed in the following section.

8.6. The Explanation of Immaterial Qualia

In general, qualia are internal qualitative impressions of what the senses deliver. According to the presented examples, it would appear that the *material basis* of qualia would be in the excitation modes of sensory perception neurons. These excitation modes, in turn, would be the neuron's inner reactions caused by the electrochemical signals from the corresponding receptor cells, which sense the external phenomena.

The material excitation modes of the sensory neurons are real and are externally observable with instruments, but also the internal immaterial appearances of qualia are real because they are our own experiences. Clearly, the material excitation modes and the immaterial qualia go together, and this leads to the question: How do the internal material excitation modes of the sensory neurons turn into the consciously perceived immaterial experiences of qualia? This is a dualistic problem, see Fig. 8.3.

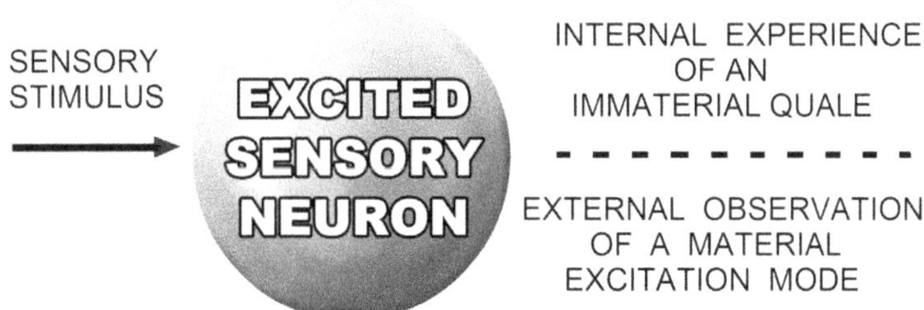

Fig. 8.3.   The excitation modes of a sensory neuron can be internally experienced as an immaterial quale and externally observed as a material excitation mode.

External observations of the excited sensory neuron do not capture any impressions of qualia. The biological workings of neuron cells can be and have been observed by various instruments like electron microscopes and probes, which can deliver imagery of the inspected neurons and also information about neural excitation modes and neuron group excitation patterns. Nevertheless, these instruments cannot detect any qualia in the way in which we experience them.

But what is the point here? What we see with our instruments are not the real things; we only get representations in the abstract forms that the instruments can produce in terms of the models provided by scientific theories. These symbolic depictions are useful and necessary for scientists and engineers, but they are not the forms of information that could be used by neurons; they are not the actual neural excitation modes.

There is no reason why the brain's neural processes should appear internally in the way of the representations that are generated by instruments. Instruments generate representations that do not belong to the category of internal experiences like those of ours.

In light of all the above, the hypothesis is: Qualia are internal experiences of the material excitation modes of sensory perception neurons and neuron groups. The internal experiences of the sensory neurons' excitation modes *are* the experienced qualia, along the ways sketched earlier in this chapter. For instance, the sensation of red is the experience of a sensory neuron's excitation mode.

Consequently, the neural location of qualia in the brain is where these sensory neurons are, not elsewhere, but this actual location is not perceived. Instead, the apparent locations of qualia are elsewhere and outside the brain, as our everyday experience shows. Neural signal patterns elsewhere in the brain do not have qualia and are not perceived at all.

But, if qualia were material neuron excitation modes, then how could they appear as immaterial? If this cannot be explained, then the explanation of qualia is not complete.

This is not a difficult question. Consider seeing a rotating wheel with spokes. We can easily see and note the rotation of the wheel as the spokes move along the rotation, and we can do this even without any consideration about the material constitution of the wheel. The rotation is material, but it is not matter. It is a condition of continuous change. Likewise, "the rotation of the wheels of cognition" is material, but not matter. However, without the underlying material "wheels", there would not be any "rotation" either.

This also applies to qualia; the qualia sensation is like the rotation, but without the material neurons, there would not be any qualia.

Qualia have immaterial appearance because the internal activity, not the matter, constitutes the experience. The material neurons and neural connections behind qualia do not enter into the equation and remain unobserved and hidden. The same also goes for the neural excitation patterns of groups of sensory neurons. Material sensory perception neurons are not observed as such, but their excitation modes are experienced. In the case of qualia, there is no need to observe material neurons at all.

8.7.   Qualia and the Observer Problem

The given explanation of qualia might be straightforward, but one essential thing seems to remain missing. Who or what would be the observer and experiencer of qualia?

I can think, and I can report my thoughts because these are a part of my mental content. Likewise, I can see what is in front of me, and I can report it because what I see is also part of my mental content. My consciously perceivable mental content is reportable.

Sensory perception produces qualia as the responses to external stimuli. Technically, the neural excitation modes of sensory perception neurons manifest themselves internally as qualia. Qualia are personal experiences, and as such, they cannot be transported, but they can be reported internally and to outside persons. I can see the blueness of the sky, and I can report this to others. They will understand what I mean if they have similar personal experiences.

What is sensorily observed can be reported. How are qualia observed then, as in the brain, there are no senses that could observe neurons and neural states? How can the mind observe qualia at all?

There is no need to observe qualia. Qualia are inner sensory experiences, and as explained in Chapter 4, inner experiences are directly part of the conscious mental content. No further act of observation is required. For instance, we do not observe our pain, we experience it.

Qualia are not observed by the mind — they are experiences and are directly a part of the mind's instantaneous reportable content.

Chapter 9

Qualia Localized and Externalized

9.1. The Perceived Location of Qualia

Localization and externalization turn qualia into direct impressions of
things and their properties out there.

Qualia are intrinsically self-explanatory, as they are taken as the things in
themselves. Red is red and sound is sound. But there is more to it.

The sensory receptor cells do not know what the stimuli are about and
where they came from. The receptor cells only react to the stimuli, get
excited and send bursts of output signals. But the receptor cells do not
know the destination of the output signals, either.

The interconnections between neurons are neither observed nor experienced. Thus, the receiving sensory neuron does not know, where the
received signals come from and what they would mean. The neuron only
experiences the resulting excitation state, which would be the experience
of a quale. That is all for the neuron, but the mere experiences of detached
qualia would not help much the actual person.

In addition to their appearance, qualia must also have an associated
property, namely their perceived location. There are bodily qualia, like the
touch and pain (if it hurts, it is me!). There are also qualia of the external
world objects. Their apparent location would be outside our body.

But then there are also qualia without any apparent material location.
These would be the qualia of our thoughts and imaginations that reside in
our mind. The mind follows us wherever we go, so do the qualia of our
thoughts. This contributes to the impression that the thoughts are ours.

Nevertheless, our thoughts and imaginations do not have any apparent material location in our head. Thoughts are just somewhere inside the head without any perceived material basis. The lack of any obvious material location of the mental content qualia contributes to the impression of an immaterial mind.

Localization and externalization take qualia out there by giving them an outside location. This is possible, as the sensory nerve fibers do not transmit the location of their input points. Therefore, the impression of the apparent origination location of the sensed can and will arise by other means.

Localized and externalized qualia are no longer just neural excitation modes appearing as qualia — they are now percepts of the body and the world.

9.2. Qualia Localized

The localization of body sensations creates the body image.

If you touch an object with your fingertip, the touch receptor in that fingertip will transmit a signal to your brain; you will feel the quale of touch and will conclude that you have touched something. But when you touch your body with your fingertip, your brain will receive two simultaneous touch sensations, one from your fingertip and another from the touched part. The conclusion: you have touched yourself. And moreover, you will know the location of the point that you have touched.

Touch receptors in the skin transmit their response to the brain via fixed nerve fibers, but they do not transmit any location information. Instead of this, the conscious sensation of touch and its location is generated in the brain.

A simple experiment verifies the above. Take a rigid stick (a pen might do) and hold it with your fingers. Next, scan a rough surface with the tip of the stick. Obviously, the stick will now vibrate a little, and this vibration is sensed by the touch receptors in your fingers. However, this is not your experience; the actual experience will be the roughness of the scanned surface. There are no nerve fibers in the stick, yet the origination point of the sensation is felt to be the tip of the stick. This would not be possible, if the nerve fiber somehow transmitted the location of the touch receptor; that location would be fixed and in this case it would be in the fingers, see Fig. 9.1. This externalization effect works also with the white cane of blind people. Also, handwriting would be rather challenging, if only the vibrations of the pen against fingers were noted, not the motion of the tip of the pen on the paper.

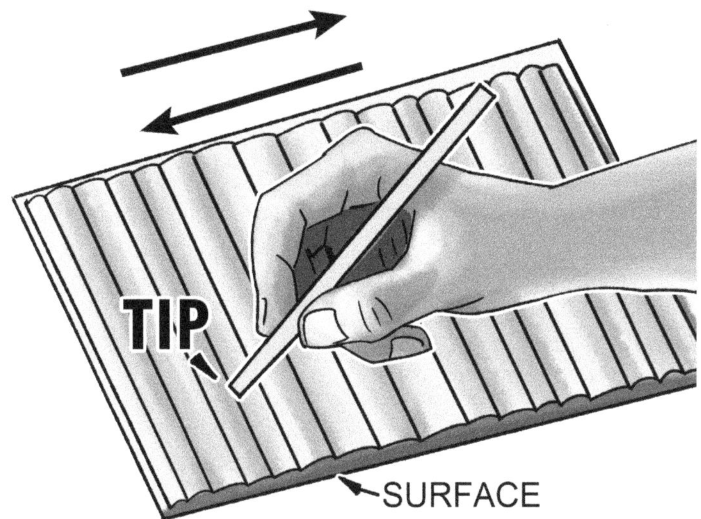

Fig. 9.1. Scanning a rough surface with a rigid stick.

Nerve fibers establish fixed connections between their endpoints like sensory receptors in the skin and neurons in the brain, but these connections do not reveal where these endpoints are. This has a consequence. We have a mental body image, we know where our body parts are, and we are able to tell the locations of any points being touched. This mental body image is not inherent or inborn. Therefore, it has to be learned — but how?

We know how it is done. For instance, sometimes it may hurt somewhere in the abdomen, but where exactly? We will find it out by touching and probing with our hand all over the abdomen, until the point where it hurts is found. Next time, when it hurts again, we will readily know where it hurts, and we are able to touch that place directly.

Little babies search their bodies by touching every part they can access. The baby sees the touched part of the body and the position of the touching hand, and in this way the touch sensation and its location can be associated with each other.

Each hand position generates muscle tension signals. These, in turn will be associated with the muscle command signals that led to that hand position.

Later on, when a body point is externally touched (for instance by a mosquito), the touch location evokes the associated hand muscle signals, and the hand can move directly to the correct position. The same mechanism applies also to vision.

In this way, the explorative touching of the body and the observing of that action will generate a mental body image. But the true fineness in this is that the actual physical locations of the relevant nerve fiber endpoints and neurons in the brain play no role here. That information about the neural structure is not used in this process, *it may and will be hidden.*

9.3. The Externalization of Auditory and Visual Qualia

Being the one behind one's eyes.

Externalization turns qualia into properties of the world with locations. This enables physical interactions between the mind and the environment. The locations of external things and the directions towards these have to be sensed, otherwise our physical interactions with the world would be very difficult.

The mind perceives the world visually and auditorily in the form of qualia. Seen things and heard sounds are out there with their perceived qualia, but these qualia in themselves do not convey direction and location information. In order to get the impression of outside existence, qualia have to be provided with locations. Qualia have to be externalized.

Without externalization, sounds would only be inside our head, just like the sound that we hear in the middle of our head, when we listen to monophonic sound with earphones. Likewise, we could see an object, but it would only be like an imagination in our mind, without an outside location. We could hear someone calling us, but we would not hear where the sound comes from.

Sound perception takes place inside our ear in the cochlea. That is the organ where sound vibrations are transformed into neural signals. However, cochlea is not the place where the sound appears to originate from. Instead of this, sounds appear to come from our environment.

It is easy to note that sounds do come from the external world. Just cover your ears, and you will no longer hear clearly the sounds of the world — the sounds are not inside your head, they come from the outside.

Sounds come from different directions, and the world around us has various objects at different locations. The directions of heard sounds and the seen locations of objects are relative to our body, which is our vantage point and the center of our perceived world wherever we go.

A simple experiment with the externalization of sound origin is presented in [Haikonen, 2019]. This sound externalization experiment is executed with a stereo headphone amplifier, good closed design headphones and two microphones. The test person places the microphones close to each other in front of the head and listens to music. Both ears will receive the same sound signal, and the perceived location of the sound will be inside the head. Next, the head is kept still and the left and right microphones are brought next to the corresponding headphone speakers. It will be noted that the perceived location of the sound is still inside the head. Then the test person turns his head with the microphones. One might think that this would not make any difference, as the common experience has it that when one listens to stereo music with headphones, the turning of the head has no effect and the sound remains inside the head. But this time there is a difference, as the microphones also move. Suddenly, the sounds are no longer inside the head, they are externalized and appear to come from distinct outside directions.

Two ears allow good sound externalization and direction detection. They also allow the association of the perceived direction with individual sounds. Sound direction detection with two ears is a comparison method. When the heard sound intensities of both ears are equal, the sound source is either in front of the hearer, or behind; slight turning of the head resolves the ambiguity. When one ear hears the sound stronger, the sound source is to the direction of that ear. Sound intensity comparison works well for higher sound frequencies, but not so well for low frequencies. At low frequencies the comparison is based on phase differences. This method is not very accurate.

The perceived direction is always relative to the direction of the head. The direction of the head, in turn, is relative to the body. These directions can be noted also in the dark because they are sensed as corresponding muscle tensions.

The ear separates different sounds from each other by their sound spectrum. Therefore, the directions of different sounds can be determined also in noisy environments.

Seeing is perceiving visually what is out there. The lens of the human eye projects an image of the seen on the retina, which is a matrix of minuscule light receptors on the retina, in the back of the eye. The retina is the place where the seen image is transformed into neural signals. The image of the seen is on the retina, but that is not our experience. We do not see the retina or the projected image on it, we see the outside world. Again, this is possible only because the visual sensory nerve fibers do not transmit the location of their input points, hence the location of the seen can and will be determined by other means.

A simple test shows that the visually seen world is out there. When you cover your eyes with your hands, you will note that you will no longer see the world — the world is behind your hands, not inside your head.

There is also another important factor in the externalization of the seen. The image on the retina is not what we scan and inspect, what we actually inspect is the outside world. We scan the world with our gaze, by turning our eyes or head. This will give us the directions towards the objects that we see, and these directions will be relative to our own position.

The turning of the head and the changing of the gaze direction will reveal the locations of the things out there, but it will also reveal the location of the optical vantage point. That point is at the eyes, and this causes an interesting impression; *the impression of being the one behind one's eyes.*

Chapter 10

The Qualia of Pain and Pleasure

10.1. The Qualitative Experiences of Pain and Pleasure

Why does pain hurt and pleasure feel pleasant?

Pain and pleasure are necessary for survival. They indicate actions that should not be done, and actions that are beneficial. They are motivators.

The subjective experiences of pain and pleasure involve mysteries. Pain and pleasure are perceived, but they are not perceived in the same way as the percepts of the external world. For example, when we see a flower, there is a flower out there, and the qualia of our percept reflect some of the properties of the flower. When we touch and feel something, there is something that we touch. But pain and pleasure are not any entities or properties to be observed, they are internal reportable experiences. Pain and pleasure are directly what they are.

Pain hurts. The feel of pain is painful and unpleasant. We want to avoid pain whenever possible. Pleasure is very much the opposite. It is the pleasant feeling that one would like to last longer, and which one would like to pursue. We want pleasure.

The essence of the experiences of pain and pleasure are their qualia. There is no mystery there; most of us know what the qualia of pain and pleasure are because we have experienced these.[1] The actual mystery of pain and pleasure is related to the neural mechanisms that produce the condition of being in pain or experiencing pleasure. These mechanisms

---

[1]There are some people with congenital disability to feel pain. This condition is known as Congenital Insensitivity to Pain (CIP) and congenital analgesia.

produce material and electrochemical excitations, while pain and pleasure are not experienced as such; the bad and good experiences of pain and pleasure are kind of immaterial. This is another appearance of the old mind–body problem.

10.2.   The Qualia of Pain

Pain hurts.

The experimental XCR-1 neural robot was first introduced in 2010. In XCR-1 pain signal is generated by hitting the robot. However, the evoked pain signal is not the pain; the "experience" of the pain itself comes from the ensuing disruptive modulation of synaptic threshold values.[2]

Pain arises from various triggering conditions. When we are injured, we feel pain. We can also feel pain, when we are ill. Inflammations produce pain. But we can also feel so-called nociplastic pain, which is pain without any clear and apparent reason.

Physical pain is related to our body, and has usually a definite experienced location. The locations of pain can be very small, like that of the prick of a needle. Larger area pain is also possible, like headaches and stomach pains.

Pain has duration. Pain can be acute, lasting from seconds to days. Pain can also be persistent and chronic, lasting for years or perhaps being life-long. Pain can be sharp and very painful, and dull with lower intensity. Sharp pain lasts usually only a short while, but dull pain may last much longer. Sharp pain is disruptive and may prevent working, while working may be possible during dull pain.

The mystery of pain does not relate much to the triggering conditions of pain, it relates to the ways in which the experience of the feel of pain arises. Modern neurology offers explanations. There are pain-sensing nociceptor cells all over and inside the body. These cells do not actually sense pain, they just respond to cell damage by transmitting neural pain signals to the brain. These signals are not the pain itself, they just trigger the generation of the qualitative experience of pain; the pain qualia.

---

[2]See the video "Robot Pain" at https://www.youtube.com/user/PenHaiko

It is also known that certain neurotransmitters are related to the pain signal transmission and experience. Among these are adenosine triphosphate, prostaglandins, histamine and glutamate. These sensitize pain nerve endings and amplify the experience of pain. There are also other neurotransmitters like GABA that weaken the feel of pain.

Pain related neural pathways and brain areas are known. It is also known how to block the neural pathways so that pain signals do not reach the brain and the pain experience is prevented. Many detailed neurological aspects of pain are known and are studied further. It might be said that modern neurology knows the mechanisms and issues of pain quite well.

However, one detail remains unknown; the subjective experience of pain. Why does pain hurt, how do the feel and qualia of pain arise? Without this knowledge our understanding of pain is incomplete. But it is usually assumed that questions involving qualia are difficult or even unanswerable and unapproachable, and therefore it would be better to leave them aside.

Nevertheless, lets approach this question from the practical experiences point of view. In the good old days, experienced electricians used to know how an electric shock feels. (Personally, I too know it well). Electric shock jolts neurons. This jolt has a specific inner appearance that some of us have learned to know so sorely. It is pain.

There are single pulse electric shocks, usually caused by direct current (DC), and then there are long lasting electric shocks, usually caused by alternating current (AC). A single electric shock is a kick that hurts, but only for a little while. A needle prick hurts in a similar way, but only weakly.

Alternating current can be considered as a series of sine-form pulses. During a continuous mains voltage AC shock, each pulse delivers its own shock. AC shock may also produce muscular cramps that prevent the withdrawal from the live electric wire or some other voltage source.

Observations like the above lead easily the hypothetical conclusion that pain consists of a single jolt or sequences of neural jolts. The jolt is what is experienced internally as the pain. The stronger the jolt, the stronger the pain.

Figures 10.1 and 10.2 present this hypothesis graphically. Figure 10.1 presents the intensity of pain versus time caused by a single shock, such as a prick of a needle.

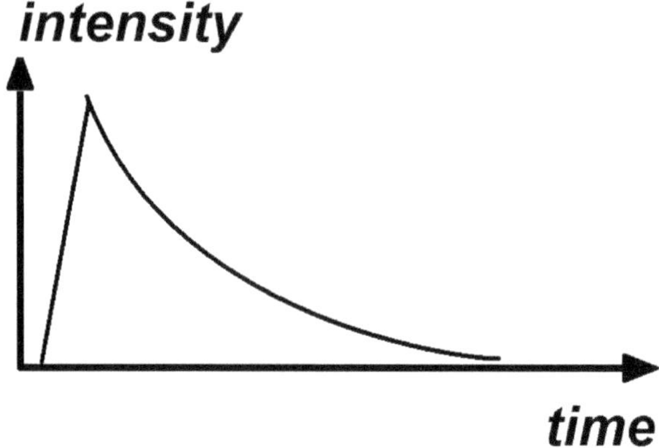

Fig. 10.1.   The intensity of the pain caused by a single shock.

The intensities of pain caused by slow and fast repetitions of shocks are depicted in Fig. 10.2. The repetition rate of the shocks affect the pain experience. When the rate is slow, individual pain peaks can be noted. When the repetition rate is fast, pain peaks will be fused.

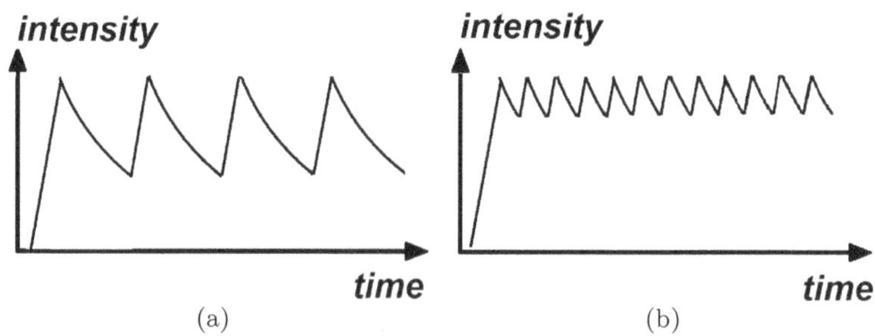

Fig. 10.2.   The intensity of pain caused by slow repetition of shocks (a) and by fast repetition of shocks (b). The repetition rate affects the experience.

To an external observer the jolt is a jolt. To the system itself it is a shock condition and experience. An external observer is not an experiencer. Observer observes what is going on, but does not have a personal experience of it; there is no feel involved.

The pain as a felt phenomenon is not shared with other sensory modalities. If it were, it would manifest itself as percepts of those modalities.

This effect does not normally happen.[3] However, pain does affect sensory perception via its effects on attention. Pain captures attention and deprives it from other activities; it is difficult to concentrate on any activities during heavy pain.

Does pain have its own sensory modality or would it be accommodated by some of the sensory modalities, like the touch sensing modality? After all, touch experiences are rather similar to pain sensations, as the sensation of a too hard touch may turn into pain.

Would this modality have a closed feedback loop? If it had, it would be prone to the runaway effect, see Chapter 15. Due to this feedback, a certain pain experience could sustain itself for long periods without any actual reason, causing *chronic pain*. Resetting the feedback loop by medical means would stop the pain, but related stimuli might initiate it again.

10.3. The Qualia of Pleasure

Sugar tastes sweet and gives us pleasure, but why?

Popular explanations for the sweetness of sugar go like this: There are sweet taste receptor cells in the taste buds on the tongue and on the roof of the mouth. The hydroxyl groups of sugar bond to these receptor cells, and the ensuing molecular responses excite the receptor cells, which then send their signals to the brain. The brain interprets these signals and produces the experience of sweetness. Similar explanations apply to saltiness and the other tastes, too.

It is also known that the neurotransmitters dopamine, endorphins, serotonin, and oxytocin are connected with the feel of pleasure.

As good as these explanations might seem to be, they still don't explain how the inner subjective experiences of tastes arise; why and how the actual sweetness, saltiness, bitterness and other tastes are felt.

First of all, there are no such chemical properties as sweetness or any other tastes to be sensed; there are only different kinds of molecules.

---

[3]This is possible. I had that kind of experience in my early childhood. Painful fever caused visual hallucinations of dark waves coming towards me, when I was lying on my bed. Later on I have not had similar experiences.

The subjective experiences of good and bad tastes are only in our minds. Here again, we face the mind–body problem, now in this form: How does the perception of different molecules produce apparently non-material impressions of properties that are not present in the sensed material molecules in the first place?

The explanation is simple; this is a trick of the mind. Qualia impressions of sweetness and other tastes do not arise from the sensed chemicals, they arise from the reactions that these chemicals cause in the inner electro-chemical environment of the sensory cells; these are neural excitation modes what are experienced, and are experienced as qualia.

A taste quale is bad, if it is connected with disruptive reactions, like in the case of pain. Bad tasting substances are more or less poisonous. Their chemical constitution is not compatible with the sensory cell's chemical environment, and causes therefore mayhem inside the cell; that is present in the cell's excitation modes.

A taste quale is good, if it is connected to soothing reactions. Good tasting chemicals fit well with the chemistry of the sensory cell. The excitation mode of the cell is smooth as long as it lasts. The ending of the experience is a disruption that may cause aggressive reactions. Parents may have noticed these kinds of aggressions, when they have refused to give more candy to a very young child.

The experiences of pain and pleasure affect attention, emotions and moods, possibly via the astrocyte neuron networks, see the following chapters.

Chapter 11

Emotions and Qualia

11.1. Emotions and Moods

The first response of a just born baby is emotional; the baby cries. From this moment on, emotions are a significant part of the growing baby's life. Emotions are the primary way in which the growing child feels about and judges the world and the situations of life. The ability to reason rationally may come later.

Emotions are important, and they do have their role in cognition. Emotional learning with pain and pleasure provides guidelines for actions in forthcoming situations [LeDoux, 1996].

Feelings and emotions have evolutionary origins. Early living cells survived by reacting to their environment by accepting or rejecting molecules. The early cells did not experience emotions, but they had states of balanced and unbalanced molecular processes.

The life of early animals was more complicated, as they were able to move around and face opportunities as well as challenges. Nevertheless, they survived by simple reactions. An encountered object was quickly evaluated and reacted to. If the object was strange, then freeze, fight or escape reaction ensued. If the object appeared to be edible, then devouring action might ensue. If the object was good for sex, then copulation might be tried. If the situation were safe for rest, then relaxation would follow. All this was based on simple stimulus-response mechanisms and related states of excitation.

The life of more advanced animals involved more complicated situations. Now the survival called for mechanisms that allowed anticipation.

Anticipation is the expectation of what might happen, and this ability can be learned by experience, if there is the capability to remember and associate things. Pain is bad, and so will also be the activity that has led to pain. The pain will be associated with the activity, which will now become avoided.

Pleasure is good, and the activities that produce pleasure are also good, and will become pursued. In this way the memories of events will be associated with goodness and badness. Remembered experiences allow anticipation of what is going to happen. This, in turn, helps to avoid past errors and allows the execution of actions that have been found to work. This has survival value.

Pain and pleasure, good and bad are also behind human emotional states. Fulfilled needs bring satisfaction. Satisfaction brings pleasure and happiness, dissatisfaction brings displeasure and frustration. Unfulfilled expectations cause disappointment and eventually despair. Good things evoke desire, bad things evoke disgust.

Some emotions are sometimes listed as opposite pairs, for example: Pleasure — displeasure, pride — shame, happiness — sadness, desire — disgust, love — hatred, hopefulness — despair.

Getting pleasure and avoiding pain are basic motivations. Introspection would seem to show that at the bottom line all other motivations were based on these. For instance, charity and selfishness would seem to be exceptions, but behind these could be the pleasure of the feel of doing something good. Or, there could be the need to appear as a good person in order to get benefits.

Emotions are triggered by events and also by memories. These are the cause for the emotions and their related expressive, physiological and behavioral reactions.

Emotions are felt via their physiological symptoms, such as elevated heart rate, sweating, paleness, nervousness, shivering, changes in breathing rate, changes in blood pressure, headache and nausea.

Emotions have also expressive reactions. The following are examples of expressive emotional reactions: Open mouth produced by a surprise, smiling when meeting a friend, laughter when something funny happens, blushing when embarrassed, crying when something bad happens, trembling voice when being emotionally shaken.

Emotional states affect also behavior. Behavioral symptoms include acts of empathy, primitive bouts of rage, retaliation and the use of expletives.

Emotions are different mental states that are responses to specific situations, events or objects. Moods are different, they are general mental states that affect our feel of well-being. Good moods make us feel happy and enthusiastic. Bad moods are related to depression, melancholy and sorrowfulness. Moods color everything and can linger on, even when the cause has been forgotten. We may feel happy or depressed for reasons that are no longer consciously explicit.

11.2. The Qualia of Emotions and Moods

It was stated in the foregoing that qualia are the way in which all our conscious mental content appears to us. There is no consciously perceived mental content without qualia. — Therefore, according to this view also emotions and moods must appear as qualia. But there is a problem: Emotions and moods are not properties of normally perceived things; they do not appear to have material origins.

Perception produces qualia that are related to the properties of the sensed objects. The color quale of a red rose is the experience of red. The quale of a square box is the form of a cube. The quale of the heard sound of repeating beats is a rhythm. Also, the inner speech is based on material processes, namely the sound patterns of the spoken speech. Likewise, our imaginations are based on what has been seen. All these qualia have material roots.

However, emotions are not physical objects with sensorily perceivable qualities. Emotions are felt — like the pain of a pain. Pain is the thing in itself, it is its own quale. In a similar way it can be suggested that the feel of an emotion is its quale. That proposition should and would do it, if in the first place, it could be explained how the feel is perceived and felt.

How is happiness felt? How does it feel to be depressed? These are tricky questions to be answered to for a reason; they invert the problem. There is no "happiness" entity to be felt. "Happiness" is just a name for an overall mental state of having sunny feelings and being relaxed without stress and burden. Likewise, there is no "depressed" entity to be felt. "Depression" is just a name for an overall mental condition that is characterized by gloominess, sadness, hopelessness and loss of interest in anything. These mental states are felt via their effects and symptoms. The same applies to all emotions generally. It might be said that emotions "borrow" their qualia from sensory percepts like good and bad taste and smell, light and darkness, weight of things. For instance: "I have a bad taste of this". "My mind is dark and heavy".

All emotions and moods have their neural basis in the dynamics of neurons and their interconnections. A special type of brain cells and their networks and connections with neurons may have a role here. These cells are called astrocytes.

Chapter 12

Astrocytes and Emotions

12.1. Neurons and Emotions

Where are emotions in the brain? If the neural signals carried meanings and neurons connected them in the associative ways outlined in the previous chapters, then apparently there would not be any mechanism left for moods and emotions. It would appear that the associative neural networks of the brain were self-sufficient and able to produce rational thinking, but not compatible with any emotional states and feelings. Simple network models of synapses and associative neuron networks would seem to amplify this impression.

Until recently, the biological synapse was seen as a two terminal device with one input and one output. This synapse passes or does not pass a chemical neurotransmitter signal depending on its threshold value. Initially the threshold value is high, nothing is passed. After learning the threshold value will apparently become lower. (There is an apparent counterpart in electronics: A diode is a two terminal device that passes current in one direction, if the voltage across the diode is higher than the diode's threshold voltage.) Figure 12.1 depicts a simple model of the two terminal synapse in comparison to a diode.

According to the simple synapse model of Fig. 12.1. There is no explicit external way for the control of the synapse's threshold value, and consequently it is only the strength of the stimulus signal itself would determine whether the signal will pass or not. That is all good, as this view has complied with what has been observed about the synapses in the brain.

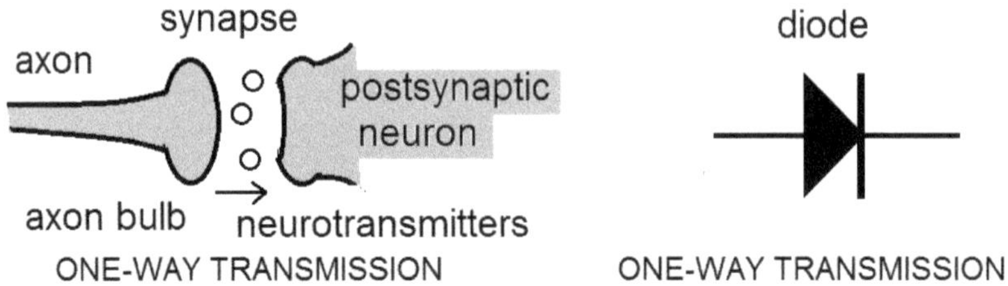

Fig. 12.1.   A simple model of the two terminal synapse vs. a diode.

However, there is a problem: This model does not dovetail well with the requirements of moods and emotions, which nevertheless seem to be an essential part of our thinking and behavior. Something more is required, and recent findings about astrocytes in the research of brain neurology may provide answers.

12.2.   Astrocytes and Emotions

The brain and the central nervous system consist of mainly two kinds of matter, which are named after their appearance as gray matter and white matter.

The gray matter of the brain consists mainly of neurons and synapses, the ones that are understood to produce perception, cognition and perhaps consciousness, while white matter has been seen only as a supporting structure that holds neurons and their networks physically together. Later on, it was found out that white matter had also many necessary "housekeeping" functions. But in recent years, advanced research has revealed that white matter seems to have an actual and important role also in the operation of the neural networks of the brain.

The white matter of the brain is not homogenous; it has structures with various cells, called glial cells. Astrocytes are starlike glial cells that are found abundantly in white matter.

Recent research has found that astrocytes establish connections with neural synapses by forming so-called tripartite synapses or three terminal synapses [Perea *et al.*, 2009; Harada *et al.*, 2016; Lefton *et al.*, 2025].

The tripartite synapse is a kind of a three terminal device, and as such it can be compared to the three terminal FET transistor, these are depicted in Fig. 12.2.

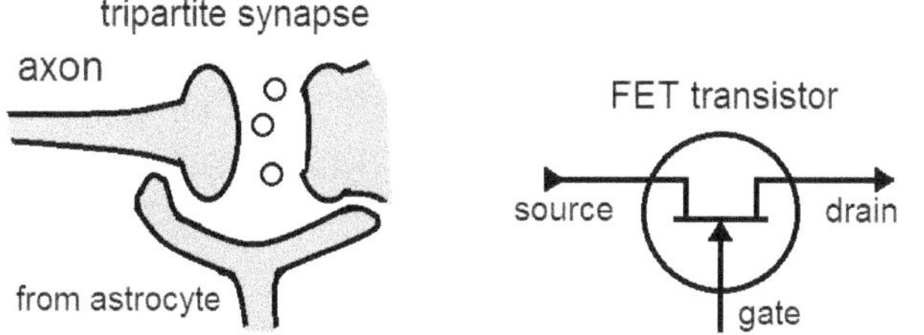

Fig. 12.2. A tripartite synapse and a three terminal FET transistor.

In electronics, a FET transistor is a three terminal device. Its terminals are called source, drain and gate. The current flows from the source to the drain, and the strength of the current can be modulated by the voltage applied to the gate. It would seem that the function of the transistor and the tripartite synapse were rather similar; namely the control of the passed signal.

In the tripartite synapse the astrocyte connection modulates the flow of neurotransmitters through the synapse. This would allow the control and modulation of the excitation level and sensitivity of the postsynaptic neuron. Thus, the flow of neurotransmitters could be disruptively controlled, see Fig. 12.3.

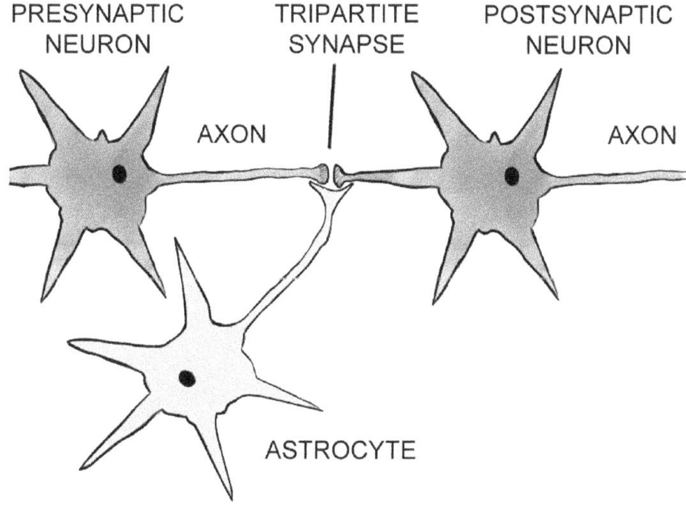

Fig. 12.3. Astrocyte connection to the synapse between the presynaptic neuron and post-synaptic neuron.

Astrocytes form connecting networks of their own. In these networks astrocytes connect with each other by exchanging chemical messages via gap junctions and also by so-called Ca^{2+} (calcium ions) waves. These waves consist of rapid increases in local calcium concentrations. These waves allow longer distance signal transmission [Scemes and Giaume, 2006].

This type of messaging allows the collective control of neuron groups. Certain neuron groups could be activated, while the activity others could be attenuated. In this way the focus of attention could be shifted between neuron groups, and the overall behavior would be modified. Different contexts would be evoked. The overall excitation and relaxation levels of neurons might relate to emotional states like fear, nervous stress and calming.

The excitation modes of sensory neurons are about sensorily or virtually perceived objects and entities, they are about the carried meanings. Moods and emotional states are not about carried meanings, they are about significance, values and overall excitation levels. These are not things with properties to be perceived as such, they become experienced via their overall excitation and slowing down effects.

Chapter 13

The Symbol Grounding Problem and Qualia

13.1. Symbols and Meanings

If you don't know what a symbol means, the symbol is meaningless to you.

Symbols are all around us. A symbol can be a character, a string of bits in computers, a gesture, a pattern, an alphabet, a spoken or printed word or anything distinct and recognizable. In effect, almost everything can be used as a symbol. Symbols are also necessary for mathematics, sciences and cognition. Without the use of symbols our modern sciences and also societies would instantly collapse.

Symbols and meanings go together. Symbols are used to indicate something; they have an intended meaning. In this way symbols can be used as "shorthand expressions" for larger chunks of information. That makes things handy, but unfortunately there is a precondition. Symbols are not explicit, as they do not inherently convey and reveal any meanings beyond their appearance. The intended meanings of symbols have to be learned, otherwise the symbols remain meaningless and useless. Thus, the meanings of symbols are not in the symbols themselves, the meanings are in the mind of the learned observer. The following case may illuminate this.

Words are symbols. They are not self-explanatory, and as such they do not transport meanings. For example, the words of an unknown language will not be understood freely, as they will not evoke any meanings in foreigners' minds because there are none. If it were not so, then we would instantly understand every language.

The problem of meaning is in fact deeper, as the following example shows. When we are reading a story, we are reading the words and sentences just like we have been taught at early childhood. Words can be read and spoken aloud, but that is not enough. We must also know what the words and sentences mean in order to understand what the text is about. But the meanings are not in the words and sentences, as words do not carry meanings. The meanings of the words have to be learned earlier and be in the reader's mind, so that they can be evoked by the read text. But that is not all, as the meanings in the reader's mind are individual and subjective. A reader will be generating its own impression of the locations and happenings of the story, as in the mind of the reader, the story is compiled by using the reader's own experiences of life and own understanding of the words. Thus, the same story may be experienced differently by different people.

Understanding a story involves the evocation of the meanings of the text, often in the form of mental imagery, like an inner movie. It would be impossible to understand the text without knowing the meanings of the words, and without any evoked inner imagery the reading experience would be incomplete.[1]

In the philosophy of mind, the connecting a symbol with its meaning is known as symbol grounding. Humans do not usually recognize that there were any problems at all there, but in pure symbol processing systems like digital computers symbol grounding is a challenging task, as only symbols can be imported, and these do not carry meanings. This makes symbol grounding difficult or perhaps even impossible. This challenge is known as the *symbol grounding problem*. The symbol grounding problem in pure symbol processing systems has not been properly solved.

13.2. The Symbol Grounding Problem in Computing

How to explain symbols with other symbols? Easy, make a list of the symbols to be explained and a corresponding list of the explaining symbols. Then, make a list of the explaining symbols and a corresponding list of the

---

[1] Aphantasia is the partial or total inability to have mental imagery. It is possible to understand text by the acquired meanings of words only, but mental imagery would help significantly. Total aphantasia may cause reading problems, and will compromise the reading experience of novels, as the visual experience would remain missing.

symbols that explain the explaining symbols. And so forth. That can be done in practice and has been done in the dictionaries. But does this really work?

Digital computers are symbol processors. They operate with digital data; binary bits and bytes, ones and zeros. That is the form of all data to be processed; every piece of information to be processed by a digital computer must be in the form of bits and strings of bits. Therefore, all analog information, like images and sound, must be digitized first by analog-to-digital converters.

The computed results are also bits and bytes, suitable for digital devices like printers, displays and process controllers. But for a human user the digital output data has to be converted into analog forms that suit the human sensory perception; visual text and imagery, also sounds.

Analog-to-digital conversion prevents the sensory experience of any direct meaning, and it also prevents the direct phenomenal connection between the meanings in the analog world and the binary data, see Fig. 13.1.

Fig. 13.1. Analog-to-digital conversion prevents the direct experience of meaning. This applies to all sensory modalities.

The computer is programmed to do calculations and other operations with binary data regardless their meanings to humans. In the computer there is no such unit, real or virtual that would actually understand anything — and how could there be; without acquired meanings there the binary numbers are just numbers, and there cannot be any understanding. This is a real problem for AI generally, and especially for the generative chatbots that communicate using natural language and vocabulary.

How to make chatbots to understand what the words mean? In principle that should not be too difficult. Just assign meanings to the words by explaining them, just like we explain the meanings of new words to our young children; words can be explained by using other words. In this way contexts can be created and new words can be explained even easier. But unfortunately, too often there is a tiny little thing that turns the possible into impossible. In this case the thing is the *dictionary problem.*

Dictionaries consist of very large number of words of a given language. The meaning of each word is explained by using other words. Also references to the external word are given by using words. When you look for the meaning of a given foreign language word, you will be given a string of words that you will not understand. No problem, just look for the meanings of these words, too. But then, you have to do it again and so forth. Eventually the word to be explained will be a part of the explanation, and you will be no wiser. Foreign languages cannot be learned only by reading dictionaries and learning the words by heart because the meanings of the words would remain missing. This is obvious to anyone who has studied a foreign language.

Ultimately, words cannot be explained by other words, and symbols cannot be explained by other symbols, there must be a ground to begin with; a small vocabulary of basic words with meanings that do not need explanation. But, alas, there are no such words, and the mere importing of words does not import meanings. Thus, pure symbol processing systems cannot acquire this kind of basic vocabulary of words with meanings, and the symbol grounding problem remains unsolvable. This has an unfortunate consequence: Digital computers cannot operate with meanings because they are symbol processors.

Without acquired meanings there cannot be any understanding, and without the ability to understand there will be no true intelligence. Chatbots will remain machine idiots that clip and paste text without any knowing about what they are doing. Sometimes they produce explicitly senseless text, and AI is said to be hallucinating. It is not. Chatbots do not utilize meanings, and therefore everything that they produce is basically meaningless. It is the user who is hallucinating when seeing some sense in the products of a chatbot. You can ask whatsoever, and the chatbot will answer whatsoever. The responsibility remains with the user.

13.3. The Symbol Grounding Problem Solved by Qualia

True reasoning and cognition operate with meanings, and therefore meanings have to be imported into cognitive systems. But unfortunately meanings cannot be imported in the form of symbols calling for explanations.

This situation leads to the obvious conclusion that importing more symbols will not rectify the situation; importing more symbols only means that there will be more symbols to be explained. Meanings have to be imported in the forms that do not require any explanation. These forms would self-explanatory information [Haikonen, 2019].

Self-explanatory information is something that is its own explanation; its meaning is explicit and does not need any external explanation. Self-explanatory information cannot be in symbolic form because the meanings of symbols are not explicit.

In human beings (and obviously in animals, too) self-explanatory information has the form of qualia — these are directly what they appear to be. Qualia present whatever is sensed, and are taken as the sensed. The quale of red is red, the quale of pain is pain. The quale of a seen square is the square pattern. In this way basic meanings can be imported into a system, if the system is able to accept and use qualia. Unfortunately, the digital computer is not such a system, it is not able to have qualia.

Qualia have also another useful property — they can be used as symbols in associative systems, like the brain. For instance, the heard quale of a spoken word may be associated with another quale of a perceived thing, and in this way the thing gets its name. As soon as this associative link has been established, the named entity can be evoked by the given name in the mind of the hearer, and vice versa, a seen entity will be able to evoke its given name. The sound pattern of the word has now become a symbol with grounded meaning.

This kind of symbol grounding is associative linking. It is useful to note that the symbol as such does not get anything in this process, neither does the linked meaning; their appearances remain the same. There will only be an associative connection between these, and any physical link doing this will not be observed by the system.

The association of meaning is not limited to the linking of two things together, it may also used to link larger assemblies of information with a simple symbol. This is beneficial as the use of languages show. Words compress information, as the actual meanings are not communicated, only the words are.

Sensory percepts have their inherent meanings in the form of qualia. But they can also have associated meanings, and this allows their use as symbols in the human thinking and cognition. Associative information processing is a method that provides tools for these tasks in artificial non-digital cognitive systems. Basic ideas of associative information processing are explained in the following.

Chapter 14

Associative Information Processing

14.1. Associative vs. Algorithmic Information Processing

The human brain is a huge neural network with billions on neurons. The exact circuit diagram of the brain is not known. And if it were known, it would not suffice. The actual operation of the brain depends on the learned conditions of the synapses, and these would not be presented in the basic circuit diagram. The situation is the same in computers. The circuit diagram of a computer does not tell what the computer does, the program code does; it is all done by programmed algorithms written by humans.

The brain is different. It does not utilize any program code written by somebody, and there is no program code to be read or extracted. Nevertheless, there have been attempts to simulate the brain or some parts of the brain by computer programs, e.g. the EU funded Human Brain Project HBP.[1]

Humanlike robots need humanlike brains. If these cannot be based on symbol processing computers with algorithmic programs, then what would be the alternative way, if any? The answer is associative neural processing, which is basically non-symbolic, non-algorithmic and not program-based.

However, humans are also able to utilize symbols and learn algorithms, and apparently there would be a case of associative versus algorithmic operation.

---

[1]HBP was a 10-year project, ending on September 30th 2023. The original advertised target was the computer simulation of the whole brain, but that was not achieved. However, the allocated 10 billion € was nevertheless successfully spent.

It is true that algorithmic symbol processing systems are not good with non-symbols due to the symbol grounding problem, while associative neural signal processing has been seen inefficient in symbol-based algorithmic computations; how exactly would you program a neural network to execute arbitrary algorithms? Apparently, there is no way, as traditional artificial neural networks are not inherently suitable for that.

Humans are not computers, yet they can learn to execute algorithms. Humans can even learn to write algorithmic programs for computers. This shows that one does not have to be a computer in order to be able to utilize algorithms. This goes for artificial neural networks, too, when they are properly designed.

Associative artificial neural networks are able to execute the transition from non-symbolic to symbolic operations and are able to naturally learn many things, including the execution of sequences [Haikonen, 2003, p. 245]. In fact, algorithmic processing can be seen as a subset of associative processing, while the other way around is awkward and difficult.

14.2.   Association in Thinking Processes

Greek philosopher Aristotle (384–322 BC) noted in his work *De Memoria Et Reminiscentia* that thoughts follow each other in an associative way; a thought evokes another because they have something in common. The same observation can be made still today.

Aristotle's association theory is known as associationism. Aristotle assumed that things became associated with each other via similarity, opposition, common motion, proximity and simultaneous appearance. Later on, associationism was largely rejected, as it was reasoned that in large associative systems everything would eventually become associated with everything in chaotic ways, making thus the whole system useless. Indeed that would happen, if context and other selection mechanisms were not used.

Here, association is seen as the fundamental operation principle in associative neural networks, and the basic operation is the linking of items with each other. The idea of associative connections between neurons was presented by Canadian psychologist Donald Hebb, who presented that "neurons wire together if they fire together" [Hebb, 1949].

In artificial associative neural networks two items are associated with each other so that later on the presence of one of them will evoke the mental presence of the associated one. The basic principle of this process is depicted in Fig. 14.1.

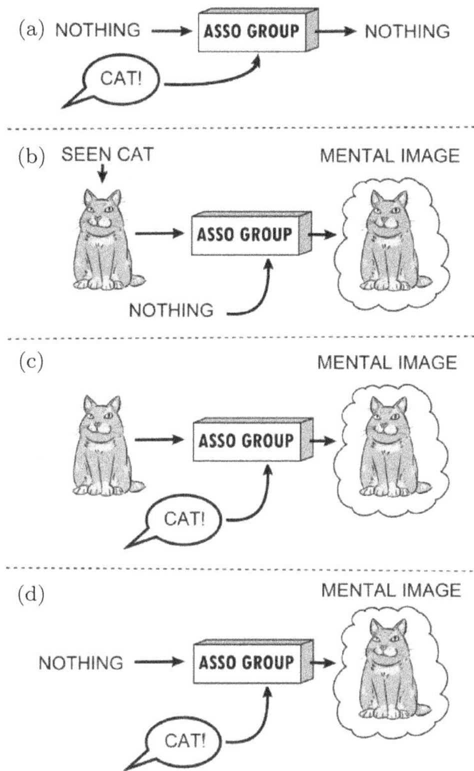

Fig. 14.1. The principle of associative linking with an associative neuron group (ASSO GROUP). (a) An unlearned asso group will not respond to any evoking stimulus ("cat"). (b) An asso group passes its input. (c) The asso group learns the associative link between its input and the evoking stimulus via repeated coincidences of these. (d) Thereafter the stimulus ("cat") will evoke the associated pattern (the mental image).

Figure 14.1 depicts a group of associative neurons (the asso group). This neuron group has two inputs. In this picture, the horizontal input accepts the neural signal pattern of a visually seen object (the cat), while the vertical input (the cue input) accepts the neural signal pattern of a spoken word ("cat"). The output, if any, of the associative neuron group is a mental image.

Initially, the vertical and horizontal neural input signal patterns are not linked together in any ways; a spoken word will not evoke any output, Fig. 14.1(a). However, any signal pattern of seen objects will go through the associative neuron group, see Fig. 14.1(b) (or may not, depending on the actual implementation).

The asso group learns to associate vertical and horizontal neural input signal patterns with each other via repeated coincidences, Fig. 14.1(c). After learning the mere horizontal input (the word "cat") will evoke the mental image of a cat, Fig. 14.1(d).

If the meanings of the horizontal and vertical inputs are exchanged, the neuron group will evoke the word "cat", when a cat is seen.

In this way a name will get an associated meaning. However, it should be noted that in this process the actual signal patterns of the object and the name will not be modified in any way; visual signals remain as visual signals, and auditory signals remain as auditory signals, and likewise for any other sensory modalities. Nothing will be added to the signal patterns, as such or in the way of any metadata.

The act of association establishes only a link between the associated items, and moreover, this link is invisible to the process in the sense that it is not observed, and the cued neurons do not get any information about the origin of the cue.

It should also be noted that the evoking neural signal patterns (the cue) do not carry any meaning. Instead of this, they only indicate the presence of the cause of the cue signal pattern. Only the presence of the evoking signal pattern matters, not what the signal pattern is about.

In artificial associative neural networks this linking is performed by special associative neurons, as explained in the following.

14.3. Neurons in Associative Processing

True associative processing is not executed as computer simulations. It is done by dedicated hardware networks, and the basic component of these networks is the associative neuron.

It is assumed that in the human brain associative processes take place mainly in the so-called association cortex areas, which take most of the cerebral surface of the human brain. The primary excitatory neuron type found in the association cortex areas is the pyramidal cell. This neuron is different from the simple neuron presented in the context of pattern recognition.

A simplified sketch of the pyramidal neuron is depicted in Fig. 14.2. The pyramidal neuron appears to have one major incoming fiber (the apical dendrite trunk), one forking output fiber (the axon) and a large number of synaptic inputs on the dendrites.

The exact details and operation of the pyramidal neuron are not yet thoroughly known, but simple models can be developed. One of these models

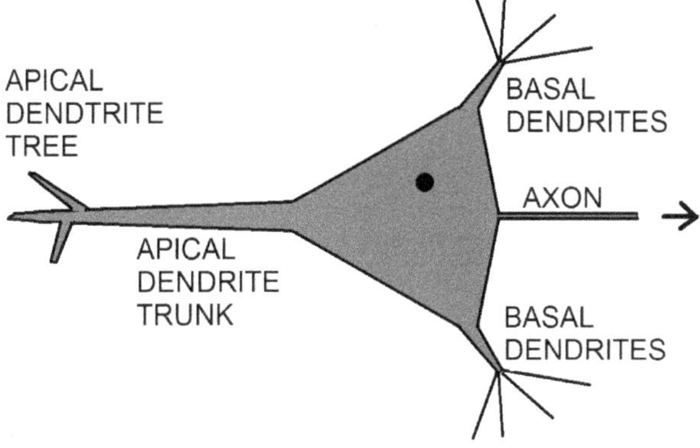

Fig. 14.2. A simplified sketch of the pyramidal neuron.

is the author's associative neuron for artificial associative neuron groups and networks [Haikonen, 2019]. The Haikonen neuron is in line with Donald Hebb's idea, but with necessary trimmings. It should be noted that the Rosenblatt [1958] probabilistic Perceptron neuron is not as such suitable for the kinds of associative neural networks presented here.

The outline of the Haikonen associative neuron model is depicted in Fig. 14.3.

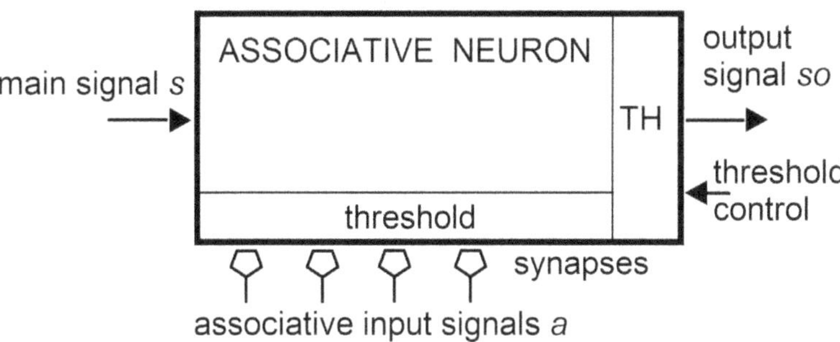

Fig. 14.3. The author's associative neuron model. The main signal and the output signal have the same meaning.

Figure 14.3 presents the outline the author's associative neuron model.[2] In this model the associative neuron would have one incoming signal (the

---

[2] See the demo videos "Secrets of the XCR-1 Robot Circuitry, Parts 1 & 2" at https://www.youtube.com/user/PenHaiko

main signal *s*), one output signal (*so*) and a number of synapses and input signals to those (associative input signals *a*, the cue). There are controllable thresholds for the input signals and for the output signal.

The neuron is designed to learn an associative link between the signal *s* and an associative signal pattern (one or more signals). This link would be learned by repeated coincidences of the main signal and the associative signal pattern. In certain applications one sustained coincidence may be made to suffice.

After the learning the same associative input signal pattern will evoke the output signal also when the main signal is not present. Main signal patterns can be handled with groups of associative neurons.

Associative neuron groups allow the transition to symbolic processing, as they can associate percept signal groups with each other; one percept signal group may be taken as the symbol or a name for another percept signal group. Afterwards this name would evoke the percept signal group of the named. The use of short symbols instead of the vast percept signal groups of the named makes processing very efficient and minimizes the number of the required neurons.

In this kind of symbolic processing the meanings are first and the symbols for these come next; the symbol grounding problem is avoided.

14.4. Serial Association and Sequences

A simple associative neuron is able to create an associative link between two things that are present simultaneously. This is good for naming things and associating properties with things, and so on. That would be quite sufficient, if it were not for one thing; the time. All things are not present simultaneously, and everything does not happen at once. Things happen sequentially, situations and episodes happen one after another; there is the present and the past, hopefully also the future. Everything changes and change has its time.

Our understanding of the world involves the ability to learn what will follow from a given present situation. This involves the association of the present situation with the previous one. However, these two situations are not present at the same time, and therefore they cannot be associated with each other in the normal way. Situations after situations form sequences, and this complicates the problem.

The solution to this problem is the use of a delay operation. The percept of what is taking place right now is temporally delayed, making it available simultaneously with the next situation. Now the percepts of the two situations can be associated with each other, and consequently a new occurrence of the first situation will evoke a memory of the following situation, see Fig. 14.4. Instead of one delayed previous situation, also several previous situations can be delayed and used. The principle of this process can be adapted to long sequences to form episodic memories. These sequences can also be named and evoked by their names, more detailed description in [Haikonen, 2007].

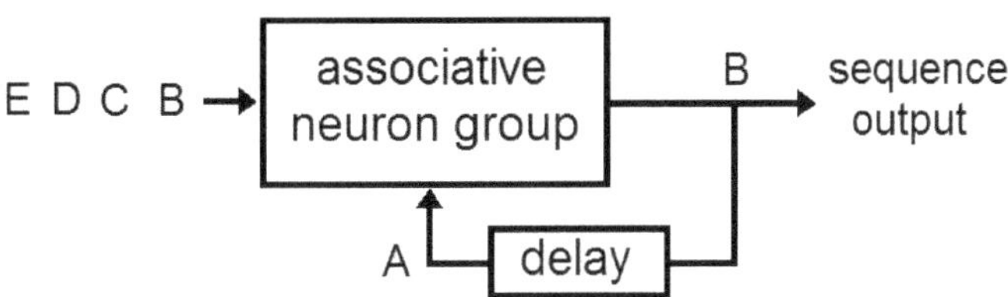

Fig. 14.4. The principle of the associative learning and prediction of the next situation in a sequence of situations by the use of a delay operation. After the learning the sequence can be replayed by inputting A. A will then evoke B, B will evoke C, and so on.

Figure 14.4 presents a system that learns the sequence A, B, C, D, E via repetitions (rote learning). During the learning B will be associated with A, C will be associated with D, etc. Now the sequence can be replayed by inputting A. A will then evoke B, B will evoke C, and so on.

The ability to predict by experience what happens next, is one application of sequence learning, but there is more. The ability to learn sequences is required in many everyday tasks, ranging from simple ones to very complicated ones. Writing a word calls for the sequential remembering of the letters of the word. Executing almost any task involves steps that have to be executed in the given order. This applies to mental tasks and physical tasks as well. These steps and their order have to be remembered, calling for sequential neural mechanisms. Walking and running are examples of these. They are usually executed consciously, while the actual sequences of the necessary neural commands remain sub-conscious.

Sequences have temporal durations, and so do their steps. Moreover, each step may have a different duration. In the sequences of physical sequences the situation is not so complicated; the physical execution of one step takes as much time as it does, and the next step will be, or can be initiated only when the previous step is completed. Then the duration of the execution of the whole sequence depends on how vigorously the individual steps are executed.

However, there are tasks that are precisely timed, such as the playing a musical piece. Here the durations of notes have specific durations, which must be reproduced exactly. The learning of these kinds of sequences calls for the learning and remembering of two things; the individual steps and their duration.

The brain is quite good in estimating and reproducing the duration of short events, as the musicians' abilities show. It is understood that in the brain there are various time-related processes taking place. However, it is not yet completely known how timing is executed in the neural level [Eagleman *et al.*, 2005].

Timing is a sub-conscious process in the brain. So, it would also be in conscious robots, and therefore the actual technical ways in which timing processes would be implemented in the robot would not necessarily matter. Many ways are available.

14.5.   Associative Attention and Selection

How to avoid chaos in associative neural networks.

The mental items in associative networks may be, and would eventually be associatively connected with many other items. That is, the activation of any of them would try to evoke all the connected items, not only the one that would be relevant at that situation. These would in turn evoke further irrelevant items, and chaos would follow. Therefore, associative networks have to utilize controlling methods that would effectively allow the evocation of the relevant item only and not the others. These controlling methods would sustain coherence and the focus of attention. Context, emotional significance and the winner-takes-all threshold may be used for that purpose.

Figure 14.5 presents the principle of selection by context in an associative neural network.

Figure 14.5 depicts a group of three neurons that are associatively connected with the same evoking signal. Here only the neuron 3 is relevant to the existing situation and its output should be selected. The selection can be executed by the use of context and the Winner-Takes-All threshold (WTA) operation. The context signal group comes from other neurons that are active in the specific situation. The neuron group 3 will be most strongly evoked and will deliver the strongest output, which will pass the WTA threshold. The operation conserves the origin of each passed signal. For technical details of the WTA and its operation see [Haikonen, 2019].

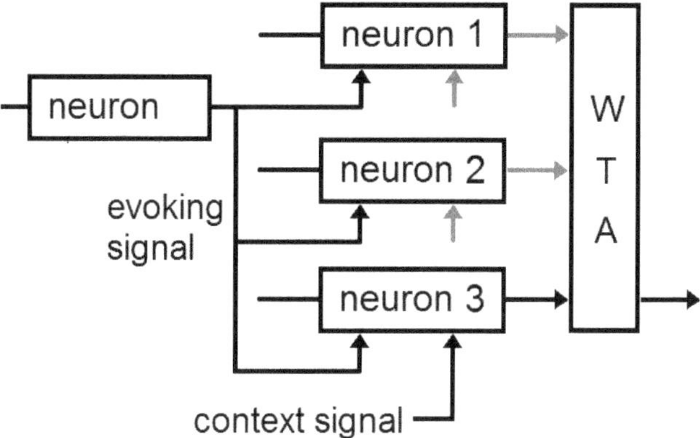

Fig. 14.5. The selection of the evoked item from many by the use of context and the Winner-Takes-All threshold.

The use of WTA allows the use of simultaneous evoking signal patterns in the determination of the correct choice of output. An example may illustrate the case. Let's assume that the human brain operates is a similar way. One may ask: What is the color of your car? In the hearer's brain the word "color" would evoke all the colors, while "your car" would evoke all the properties of your car, including its color. One of the colors would be evoked doubly, letting it to pass the WTA. The correct answer would be produced. This neural process would remain subconscious, while producing the correct answer.

Similar mechanisms can be used when attention and selection control is executed by emotional significance.

Chapter 15

Perception, Feedback and Imagination

15.1. Sensory Perception with Feedback

Humans have many senses, which produce different percepts of the external world and the body itself. Eyes produce visual percepts, ears produce sound percepts, touch sense produces touch percepts, and so on; they all have different qualities, the qualia. In the brain, all these percepts are processed by a limited number of different types of neurons. The question is: Do these various sensory systems have completely different neural networks for the processing of their own kind of information, or are there some common network architectures? This question is relevant to neuroscientists and also to robot brain designers.

The use of perception-response feedback loops in sensory perception neural networks is based on the assumption that there is a common basic neural architecture for all sensory modalities, possibly with certain modifications to comply with the specific nature of each sensory modality. This assumption is based on the fact that regardless of the differences in the handled information, there are a number of general, common requirements for the perception processes of each sensory modality and also for the interaction between these modalities.

The idea of feedback or re-entry in the brain as a sketch is quite old. However, advances in brain research have enabled more concrete ideas of neural feedback systems. The benefits of feedback in perception processes have been recognized e.g. by Chella [2008] and Hesslow [2002]. The author has presented a perception-response feedback loop architecture in [Haikonen, 1999, 2003, 2007, 2012, 2019].

Senses produce sensory experiences. There are several sensory modalities, like vision, auditory, smell, taste, pain and body part position (proprioception). The general sensory perception process can be depicted as in Fig. 15.1.

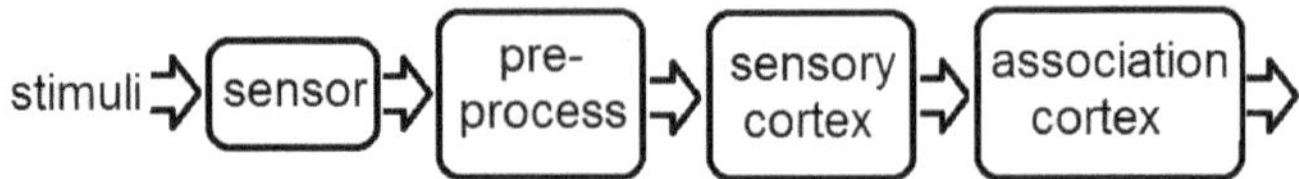

Fig. 15.1. The general outline of the sensory perception process.

In Fig. 15.1, the sensor receives stimuli from the environment, and produces responses to that. These responses are pre-processed and delivered to the sensory cortex in the brain. There, the percepts will have the appearance of qualia. Qualia-related neural signals are then delivered to associative neuron groups in the association cortex of the brain, where they lead to the production of the system's final response.

Sensing is pointless without the production of some kind of response. Possible reactive reactions aside, the responses of each sensory modality are of the kind of what is sensed. The responses of visual modality are mental imagery, the responses of the auditory modality are inner sounds, and so on. That would be all good, and would serve useful purposes, if there were somebody watching, listening and feeling. But the seeing of the produced inner visual imagery, hearing inner sounds, and so on would require inner sensors for these to perceive them — and to produce further percepts to be perceived with further sensors, and so on, perhaps endlessly. We have big brains, but not that big to accommodate all this.

Luckily, these chains of inner perceivers and observers, or "homunculi" are not required, as the required perception processes are there already. They are the sensory modalities themselves, and by using feedback they can be reused to perceive also mental content.

Feedback loops solve the problem of the perception of inner mental content by feeding the produced responses back to the sensory perception process to be perceived as virtual sensory percepts, see Fig. 15.2.

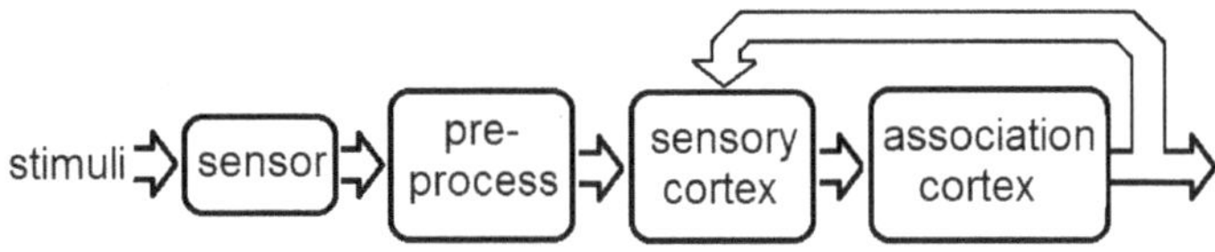

Fig. 15.2. The perception-response feedback loop.

In Fig. 15.2, the feedback loop sends back the response from the associative neuron group to the perception process via a group of feedback neurons. It should be noted that the feedback is associative; the sensory cortex neuron group does not receive anything via the feedback loop; the feedback signals only evoke pre-process signals even when these are not produced by the sensory pre-process.

In this context it should be noted that feedback loops may be tricky; they may suffer locking and runaway effects. In the locking condition the feedback signal will be circulated indefinitely, and in the runaway condition the feedback information varies in rather random ways, but still sustains itself. A designer of these kinds of feedback loops should be aware of these possibilities.

Feedback allows the perception of the mental responses of each sensory modality by the modality itself. This has further benefits. For instance, an object that has been seen earlier, will evoke an inner image (a mental sketch) of the same. When this "image" is fed back to the feedback neuron group, it will evoke a virtual percept, which will have common features with the sensed "image". The actual sensed "image" and the evoked "image" will then match at feature level (not "pixel-by-pixel" level!). The matched feature signals will be doubly excited and will thus be amplified, and the object is experienced as familiar.

In a similar way, when an object is searched, there will be an expectation, a mental visual sketch of what is to be searched for. When the seen object matches the sketch, the searched object has been found.

Feedback allows the determination of match/mismatch and novelty (m/mm/n) conditions of the sensed percept. But it will allow and facilitate even more, like imagination and the continuous inner speech.

Feedback is an essential function in every sensory modality. However, feedback alone cannot provide all that is required for cognition. It cannot give its percepts any meanings that are outside the modality, as the meanings would be percepts of the other modality. Meanings arise from associative connections, and for that purpose the perception-response feedback loop modalities must be associatively cross-connected. These cross-connections facilitate the making of sense of what is perceived sensorily and virtually.

15.2. Cross-Connected Sensory Modalities

Sensory modalities produce sensory information of their own kind and with their own kind of qualia. Together, this information should produce to the mind the moment-to-moment connected experience about the environment.

This connecting would, for example, allow the motor modality to make an approach toward a seen object and possibly grab it. However, these coordinated actions will not happen, if there are no connections between the modalities.

At first sight, it would appear that it is the mind who perceives what the senses deliver, and will form an overall view of the situation. The conclusions and physical actions would then be based on this view. After all, don't we see and hear what is around us, and don't we decide to take actions as necessary? If, for instance, if we see that there is a book on the table, we can decide in our mind to pick it up, and so we do. The mind sees, perceives, connects and decides. This should be obvious.

In the foregoing it was argued that each sensory modality is its own perceiver, there are no additional unit or units for that. Along the same line of reasoning, there would not be any additional units for some kind of sensory information integration. There would not be a mind-entity.

The sensory modalities have to be cross-connected, but there is a problem. Sensory information from one modality cannot be transmitted as such to the other modalities, as there it would only cause havoc. Sounds could appear as visions, smells as sounds and so on; chaos would ensue. The situation is: nothing cannot be transmitted or received, yet the various sensory modalities should be connected with each other.

Associative linking provides the answer. When two items are associatively linked, then the presence of one item evokes also the other one. Both of the linked items remain the same, they do not receive anything from each other. The link between the items is the only established thing, and even the link is invisible to the linked items.

Associative linking can be easily implemented in associative neural networks. In fact, in these networks associative links may automatically arise by themselves via repeating occurrences of simultaneous presences of the items to be linked; there is no need for any advance planning.

Figure 15.3 depicts the principle of the associative linking of two perception-response feedback loop sensory modalities. The sensory cortexes of the modalities hold their percept, which is at the focus of momentous attention. Each modality sends their percept signal patterns to the associative neuron groups of the opposite modality. There the received signal patterns are not interpreted or received as such. In fact nothing will happen, if these percepts have not appeared simultaneously earlier and in this way have not generated an associative link. If the link has been

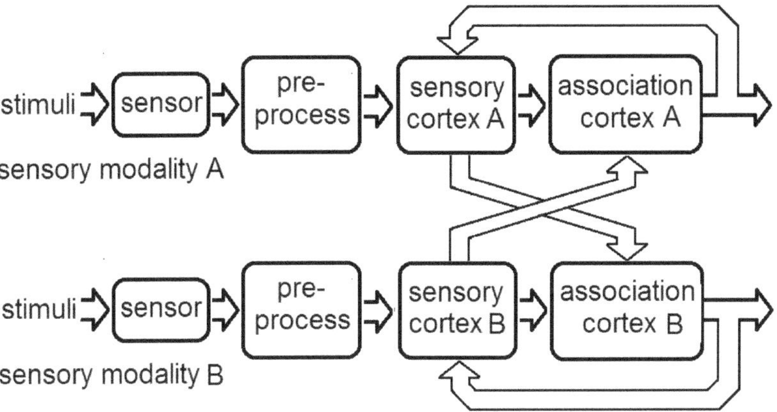

Fig. 15.3. Associatively cross-connected perception-response feedback loop modalities.

established, the received percept signal pattern will evoke the linked percept. For example, in this way a spoken word, a name, may evoke an inner image of the named object, and vice versa.

It is useful to notice that the sending modality does not know whereto its output goes, and likewise, the receiving modality does not know where the received signal comes from. And moreover, the receiving modality does not know what the received signal pattern means. The received signal pattern merely evokes the associated percept, and this percept just appears to come out of nothing.

Similar effect can be noticed by introspecting one's overt and inner speech; words and sentences just pop out without any hint of any machinery behind the process. The philosophical immateriality aspect of thinking manifests itself there, and in this way has its explanation.

Many sensory modalities may be cross-connected in this way, and this creates a pre-condition for cognition by associative processing.

15.3. Imagination

Feedback loops facilitate imagination.

Imagination is the ability to generate mental visions of something that is not actually present. Imagination is not only for scientists, inventors, creative novelists and artists; it is for everybody, and it is utilized in everyday situations. Our plans for the rest of today and the forthcoming days, weeks and years are just imaginations.

Imagination is not remembering, but it utilizes memories of what has been experienced; visual shapes, sounds, actions, situations. These are the building blocks for imaginations. The building blocks can be rearranged, modified and utilized also in ways that do not or cannot happen in reality.

The seen and otherwise perceived things have different features. Things have different shapes, sizes, colors and so on. A given object may have a certain form, while it can appear in different sizes, yet it is the same object, and should be recognized as such. The world has immense amount of objects, and each of these objects rarely look exactly the same, when perceived again. That would be an insurmountable challenge for recognition, unless there were a method to overcome it; namely the detection and use of invariant features. A circle is a circle regardless of its size, a square is always a square, and so on. Objects are recognized by their invariant features, which actually are just few. This is verified by a common observation: a simple sketch is able to convoy the impression of a given person. Cartoons verify this.

The use of features solves the recognition problem, but it also provides freedoms for imagination; imagined objects can be modified by simply adjusting their features.

Mental imaginations are perceived via feedback, and are like vague products of perception, however about things that are not there. Imaginations are generated, modified and perceived in terms of what our senses have delivered to us. It is possible to imagine things that do exist and do not exist, but it is difficult to imagine something that cannot be presented in the terms of any of our sensory faculties.

Imaginations do not have to be executable in real life. This fact is well-known in Hollywood.

Imaginations may also be emotionally weighted. Daydreaming of something nice and pleasant will give pleasure, while imagining some possible personal misfortune will not.

Imagination facilitates planning. Our conscious actions are planned, and are based on what the situation demands or offers to us. For example; I feel thirsty, and I know that there is a soft drink bottle in the refrigerator. I can imagine myself going there and taking that bottle. That imagined action constitutes my plan.

Imagination is also a necessary tool for engineers and architects, as it is the first step in designing anything. The wonderful process from an immaterial idea to a material end product would not be possible without the faculty of imagination.

As processes, imagination and conscious sensory perception are similar. One can only imagine something when one is conscious; imaginations are reportable virtual percepts, which are facilitated by the internal feedback loops.

Also nocturnal dreams are imaginations produced subconsciously, yet they are perceived consciously and can be reported, if the dreamer wakes up quickly after the dream. But during sleep this conscious perception is not perfect; actual sensors are cut off, and will not provide reality checks.

Imaginations are perceived as some kind of virtual sensory percepts, but there is a difference: Our thoughts and imaginations are normally inside our head, while the true percepts of the outside world are externalized; they have perceived locations out there, outside of our head. However, it is possible to lay imagined objects on top of actually seen objects, and it is possible to imagine actions and motions, which the real seen objects could execute.

15.4. Imagination Augmenting Perception

At each moment, the contents of our consciousness consist of sensory percepts, inner speech and inner imagery; the imaginations. It turns out that imagination is necessarily connected to sensory perception.

The senses facilitate perception, and perception produces the experience of being in the middle of the external world. The seen objects, the heard sounds, smells and situations are real, and they are out there, around us. That is all good, as without this sensory experience our life would be difficult. Yet, this is not sufficient.

Sensory perception provides information of what is present right now. Memories provide information about what was before. Imaginations provide ideas of what is to come in the future. We see and hear that something is happening, and try to understand it in order to act properly. But perception does not reveal everything and memories are imperfect. What is partly or totally hidden, what is going to be and what we should do next and in the future, can only be imagined and expected.

Expectations modify perception. We see what we expect to see, we hear what we expect (or want to) to hear. And not only that; when we see something happening and hear something being said, we will have expectations about what will happen next and what we will hear next. Some expectations are subconscious, while some are consciously produced. Our percepts are interpreted within the framework of our expectations, and are laced with our imaginations.

Imagination contributes also to situational awareness and mental models of world.

Chapter 16

Memory in Computers and in the Mind

16.1. Computer Memories

Humans have vivid memories of past events and experiences. Good memories are reminisced with pleasure, bad memories are not. These memories are accessible by their content; a cue will evoke a specific memory.

Computers do not have memories to be reminisced in the human way. Digital computers have only binary data, which is stored in computer memory circuits.

In computer memory circuits data is stored in addressable memory locations in the form of sequences of ones and zeros. The storage and recall of a specific piece of information requires the use of the corresponding memory address, another sequence of ones and zeros.

Databases are vast organized collections of data in computer memories. But the very size of these databases amplifies the search problem. How to access specific data, what is where, how to know the correct memory address; these are the data search problems in search engines. Metadata solves these problems, but it has to be generated first.

Metadata is data about the data itself, about its topic and location. For information retrieval purposes contemporary metadata systems utilize so-called HTML title tags. These are small "labels" for different data topics. Search machines utilize these tags to find data about the searched topics. Consequently, all data contents must be tagged first with these tags to make this kind of search possible. For historical reasons there used to be lots of non-tagged information in the web. Going through it all and tagging means a lot of work.

Content addressable memories would solve this problem. In these memories pieces of the content would work as the memory addresses directly. The principle is simple, but the making it to work is not.

There are two kinds of content addressable memories, namely auto-associative memories and hetero-associative memories.

In auto-associative memories the content is retrieved by a piece of the memory content, in hetero-associative memories the content is retrieved by some associated data, for example, a name.

A simple content addressable memory can be realized by using a conventional digital memory chip with separate input and output lines for data, and using the cue data as the address input, see Fig. 16.1.

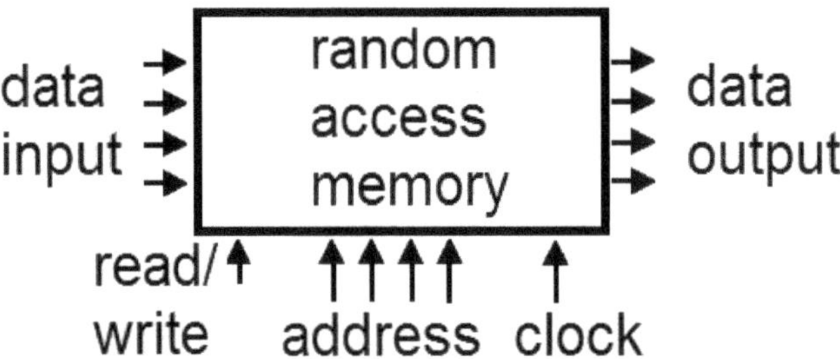

Fig. 16.1.   Random access digital memory with data inputs, outputs and address input.

During memorization the address (the cue data) and the input data are provided at the same time and the write command is given. During recall the address is provided again, and the read command is given. The memorized data will then appear as the output.

This simple content addressable memory has two shortcomings. Firstly, the content is associated with the cue data in one take. Thus this memory cannot be used in correlative learning in which many learning examples are required. Secondly, the cue data must always be exactly the one and same, otherwise the correct content will not be retrieved. In practice, however, the same content should be accessible also with almost similar evoking information. Thus the address data would not be exactly fixed, and there would be small variations to be tolerated. Conventional memory chips do not allow this. These are the reasons why computer memory chips cannot be readily used in simulations or emulations of biological neural networks.

In contrast, the human brain is an efficient neural search engine. Its neural processes are very, very slow in comparison to the execution speech of computers, yet it is able to beat the computer many times over in memory search operation. In healthy persons memory recall is instant and effortless, especially in the production of speech.

16.2. Memories in the Mind

Personal memories give us personal history. Accumulating memories will also create the impression of the flow of time.

If one cannot make memories of anything, one will not learn anything and will not acquire experiences and knowledge. Moreover, without the ability to store memories of the past, one will not be able to accumulate any personal history; one will not know who one is. And this is not all. Situation awareness is not possible without the ability to remember recent events and thoughts. Long-term and short-term memories make one into a functioning individual.

The brain has various memory functions, which are usually classified as sensory memories, working memories, short-term memories and long-term memories.

Each sensory modality has a short specific memory, which sustains instantaneous percepts for a while. This is necessary for the detection of change via the comparison of the present percept with the previous one. Without this operation percepts would not have detectable durations; each percept would be kind of timeless, as the previous percept would not be available. Also, the ability to sustain percepts has other benefits.

Pain is sustained for a good while after the cause of pain has vanished. The sufferer will understand that some kind of first aid should be taken and will also learn to avoid the pain causing action in the future.

Auditory memory allows very accurate recall of a short auditory experience. A heard sentence or a heard short sound may linger on as if these were heard again. A person may ask "what did you say", and at the same time "hear" again what was just said. A bang may echo in the mind for a while with all its auditory qualities.

Eidetic memory allows the sustaining of the seen for a while. Adults do not usually have much in the way of this ability, while young children may have. Eidetic memory should not be confused with negative afterimages and imaginations.

Working memory allows the remembering of the items of a thought. This memory facilitates the continuity of thoughts, but it has a very limited capacity; only past few thoughts can be effortlessly recalled.

Short-term memory is an episodic memory. It allows the recall of recent episodes; what one has done today and yesterday, perhaps last week. The details will soon fade away, as new memories replace the old ones. Short-term memory allows the evaluation of the emotional significance of the past moments, marking these memories worth saving. Short-term memories have temporal continuity; the temporal order of the episodes is preserved.

The structures in the brain that are responsible for these memories, consist of two similar parts called hippocampus, together they are called hippocampi. It has been experimentally (accidentally) noted that the removal of hippocampi leads to the inability to make short-term memories and new long-term memories [Squire, 2009]. The patient will not be able to remember what happened few minutes ago or earlier, and will not accumulate personal history. Moreover, the patient will not experience the flow of time; personal time will be frozen.

Long-term memory has a very large capacity, and may store memory snapshots from early childhood to the present. These memories are not necessarily very accurate, and their temporal order may be distorted.

16.3. Making and Recalling Memories

Where are long-term memories located in the brain?

When notes are written down on a notebook, permanent physical marks are made on the pages of the notebook. Likewise, in the brain the imprinting of a memory involves the making of permanent physical changes in the neural networks of the brain. But marks on notebooks are nothing in themselves; they have to be seen, read and interpreted.

The imprinting of the memory trace is just the first step, the recall is another. Memories have to be recallable, and the recalled neural patterns must be returned into consciously perceived percepts in the terms of the original sensory experience; this process must allow the "reliving" of the recalled memory. Introspection shows that recalled memories have the form of qualia, even though these qualia may not be as complete and vivid as those of the original event.

The recall of memories calls for a mechanism that facilitates the recall of the relevant memory at will. Considering the vast amount of memories in the brain, the brain's recall mechanism is very fast and efficient, yet subconscious. Vastly parallel associative recall would seem to enable this.

Where are memories stored in the brain? Vinyl and CD records, tape recorders and digital recorders are artificial memory devices; they are able to store and replay sounds and music. The stored information has a definite material location like the groove on the vinyl record, the magnetic variations on the tape recorder tape and the numeric values in the addressed memory locations in digital memory chips.

It might be assumed that the same would apply to the human mind; the locations of memories could be found in the brain. But is it so? The author's hypothesis is: *Long-term memories are not stored anywhere in the brain*. Obviously, this claim needs some explaining.

Neurons are connected with each other via synapses. During learning the synaptic transmission strength increases, and the pre-synaptic and post-synaptic neurons will have an associative link with each other. Indeed, it has been assumed that synapses are related to memory making. But synapses just connect, they do not remember things and events. The transmission strengths of synapses are not descriptions of anything; they only facilitate the evoking of the output signal of the neuron.

Nevertheless, it has been presented that the synapses enable the creation of new connections and neural circuits that assumedly represent and store memories. But this assumption as such is not a complete explanation. How would a neural circuit store and represent a memory? How would this neural circuit be observed as a memory? Our common experience is that we are not able to observe our neural circuits. The represented memory should be observed, but how would that happen?

The author presents that the there are no representational memories there to be observed, there are only synaptic connections that allow the reconstruction of the memories.

Our recalled memories have the form of inner imagery and verbal descriptions by the inner speech. These, in turn, have the form of sensorily produced qualia. Only sensory perception operates with qualia, and therefore recalled memories have to be reconstructed at the sensory perception points. Therefore there has to be a mechanism in the form of feedback loops, which return the evoked memory signals into the perception process to be experienced as virtual percepts.

Figure 16.2 presents the feedback loop model in memory recall. During memory making, sensory stimuli are sensed and the sensory signal patterns are passed through the percept point to the association network. (Qualia are not transmitted, they are local experiences.) There they are associated with other active signal patterns, which relate to the episode being memorized.

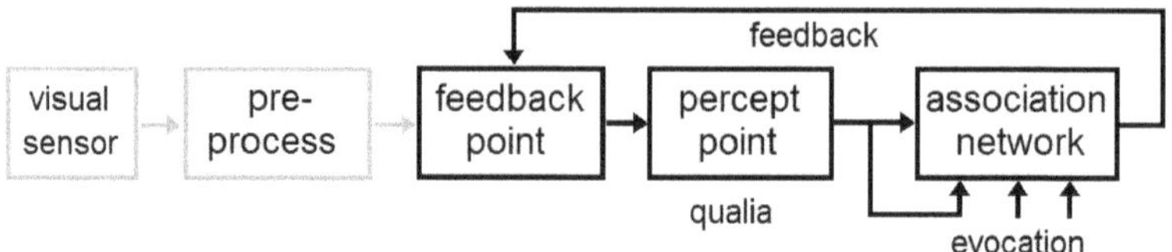

Fig. 16.2.   The feedback loop in the reconstruction of visual memories with qualia.

The memory will now be a latent complex of associatively connected neurons. Later on, when a small part of the complex is evoked by some cue, the whole complex will be activated. The mere activation does not produce consciously perceived memories; for that to happen, the evoked signal patterns will have to be transmitted back to the feedback point, where they will excite the corresponding qualia-related neurons with their neural excitation modes. In this way memories will be reconstructed in the terms of sensory perception with qualia. These memories do not exist anywhere to be recalled directly as representations, they are associative reconstructions.

The feedback loop model explains how memories are evoked and reconstructed and how they will be consciously perceived. There seems be some actual proof for this model; it is has been noted that the visual sensory cortex is activated during the imagination and recall of seen visual objects [e.g. Dijkstra, 2024].

Models aside; the brain is not a tape recorder, and memories are not recalled as such; they are reconstructions. The reconstruction process is not perfect, and memories are not accurate also when they appear to be perfect. Also false memories can be experienced and created.

The feedback loop architecture and the associative neuron groups can be extended to handle also episodes. The feedback loop is also essential to inner speech, allowing the associative evocation of the next words and sentences, and in this way the continuous flow of thoughts.

16.4.   Time, Memories and Personal History

Without memories there would be no time, only the present moment.

Time goes by. Things happen and we get older. We can note this without any clocks and mirrors. And how are we able to do this? — We have memories of the past, and we can remember.

In general, it can be noted that time is only a concept. When nothing changes and also clocks stand still, there is no way to observe the passage of time. If this were a persistent situation, the concept of time would be useless. But in our universe many things happen simultaneously and sequentially. Yet, there is no actual flow of physical or non-physical anything that could be called time. There is only change; the events. "Time" is created by changes. The non-existence of time has been proposed e.g. by physicist Ernst Mach (1838–1916). See also Barbour [1999] and Haikonen [2023].

Clocks provide "ticks", series of events, and allow the counting of these ticks between the beginning and the end of given events; the events get a measurable duration. Time measurements are comparisons between the change rates of events.

The human body has its own "clocks" and rhythms, like the heart rate, breathing rate, metabolism and the brain wave rates.

Each moment of time consists of a snapshot of the changing material and energetic order. As physical beings we are a small part of this order; we are not living in time, we are a part of it [Haikonen, 2023]. (This has implications to time travel; we would not be able to travel back to any given moment of past, because our presence there would change the past into a different one. There are also other reasons, like the energy conservation principle, which makes time travel impossible.)

The world provides yardsticks for time. The Earth orbits the Sun, and seasons follow each other. There will be winters, springs, summers and falls. The Earth rotates, and days follow each other. There will be nights, mornings, noons, afternoons, evenings. Our lives are rhythmed by these. There will be time to work, time to rest. There will be time to be young, time to be old. There will time to be alive and enough time to be dead.

Humans are able to observe these changes, and consequently will acquire a concept of the passing of time. Here, the ability to form short-term and long-term memories has a decisive role.

We can remember what happened just now, what happened recently and what happened a long time ago. In our minds time goes by because events follow each other and our memories accumulate. And time goes forward, as there will be more and more memories. If this did not happen, we would only have the present moment.

How long is the experienced present, then? If nothing changes, the mental unchanging present continues. Time passes by slowly for those who wait for something to happen. On the other hand, if there were rapid chances, the moments of the present would be short, and the time would seem to fly.

We have the remembered past, the observed present and the imagined future. We can remember what has happened in the past in general, but memories are also our own personal histories. Our personal memories shape contribute to our self-image.

Conscious robots should also acquire the concept of time and their personal history in a similar way. For this purpose robots should have short-term and long-term memories. These in turn, would facilitate situational awareness in robots, see the following.

Chapter 17

Situational Awareness and Mental Models of the Word

17.1. Situational Awareness

Crossing a busy street may be dangerous, if you do not have situational awareness.

Open eyes enable seeing. But mere seeing is not seeing of what happens. Even the seeing of what happens is not enough. You have to understand the situation, and foresee what is going to happen next. Without situational awareness you are at loss.

Where am I now? Why am I here? What am I supposed to do now? What have I already done? Persons with failing situational awareness cannot answer these questions. They may not recognize familiar places and may not remember how they have ended up there and for what purpose. The loss of situational awareness can be a scary and confusing experience, and potentially dangerous, too.

Situational awareness relates to current situations and requires conscious perception of one's environment, actions and aims. If you are not conscious, there will be no situational awareness, either. But being conscious is not necessarily same as being situationally aware; one can be conscious, but confused.

Conscious perception delivers continuous information about the circumstances around us. But this alone does not suffice; we must also be aware of our own intentions and actions; what we are doing, and how this fits with the environmental situation. We must remember things, and for this short-term and long-term memories are required for this.

105

The ability to continuously form and utilize short-term memories is necessary for situational awareness. Without properly working short-term memories one cannot remember what happened just a moment ago, and there will be no perceived moment-to-moment continuity. Without this continuity each moment will be just an isolated experience without any connections with previous moments; the situational context is lost. "Where am I, and why" experiences may follow. These kinds of experiences are seen in demented people with damaged hippocampus. Also temporary confusion may also cause a short loss of situational awareness. One can be conscious without situational awareness, but not the other way around.

Effective situational awareness is predictive, and it will create expectations of what will come next. These expectations are produced by earlier experiences of similar situations. These expectations may also be subconscious and normally unnoticed, but they still exist and operate in the background. The presence and effect of these kinds of expectations will become noted only when they are not met.

The empty milk carton effect is one example of this: You are reaching out for a milk carton on the table to lift it up. You do it, and the carton and your hand shoot up high. The carton was not full and heavy, it was empty and light. Your mind had subconsciously commanded the execution of the act with the expected amount of force required for the lifting of a full carton. This subconscious expectation was not met, and surprise ensued. At that moment the presence of the subconscious expectation about the weight of the carton revealed itself.

The conscious and subconscious expectations change as the situation changes. As being predictive, expectations are and should be a step ahead of what is going on. Situational awareness is faultless as long as the flow of expectations matches with reality. Everything is all right as long as this condition continues. One is the master of the situation, and there is no need to even note and pay conscious attention to how the situation evolves.

However, things would change, if the reality suddenly deviated from what was expected. A mismatch condition between the expectation and the reality would follow, and this mismatch would cause alarm. Attention focus would be quickly shifted to the changed situation, and corrective actions might be taken.

Situational awareness is about observing, understanding and foreseeing. It is about coping in everyday life. Simple mental models of the word help there, but extensive mental models can do even more.

17.2. Mental Models of the World

We live in our imagined world.

We all know that the world operates in certain predictable ways. Days are followed by nights, winters will come after summers. Things fall down to the ground, fire burns, without money you cannot buy anything. There are causes and consequences. Observations like these are examples of simple wisdoms, created by everyday experiences beginning from our early childhood. These are part of our mental models of the world, against which we match and evaluate consciously and subconsciously our daily situations.

Mental models of the world are personal, subjective frameworks that allow the putting of things into real or assumed context, and in this way they help to understand and manage situations. Good mental world models make the perceived world coherent and predictable, and this allows meaningful actions and the planning of future.

Different personal mental models of the world exist. Spatial mental models are "maps" of our home, our neighborhood, and other familiar places that we have visited. Temporal mental models are related to the passage of time and its effect on our actions. Long-term memories are related to these.

Then there are also general mental models for causations, connections and interactions; how people interact with each other, how things interact with each other, how occurrences interact with each other, and what kind of consequences these interactions might have.

Political and other opinions and worldviews are based on these personal mental models. We have our own bias on these, our worldviews and opinions are subjective.

In societies, common worldviews are useful as they make people's behavior predictable. That happens regardless of the validity of the worldview.

Models are just models, not the real thing; they are not accurate representations of reality. They are just tools. This applies also to our mental models and views of the world. At the end of the day, our personal worldviews and mental models of the world are only subjective, and are pretty much created by subconscious imagination to suit our subjective worldviews.

We are not able to import the real world, as our senses can only deliver their reactions to the real world's phenomena. It is up to our mind to make sense of the world with the help of mental models and frames. Ultimately, this is done by our imagination; we live in our own imagined world. — That would seem to explain many things.

All this is essential also to humanoid robots with humanlike behavior. They also need situational awareness and mental models to execute their assigned tasks and to copy with the changing circumstances. But there will be no situational awareness, if the robot is not conscious.

Consciousness seems to be necessary for humans and humanoid robots. If so, what is this consciousness, actually? See the next chapter, the final conclusion may be a surprise.

Chapter 18

Consciousness Explained

18.1.　The Mystery of Consciousness

All attempted explanations of consciousness have failed, and there is a simple reason to that.

Anil Seth, an eminent professor at the University of Sussex, has presented: "Consciousness is, for each of us, all there is: the world, the self, everything. But consciousness is also subjective and difficult to define. The closest we have to a consensus definition is that consciousness is 'something it is like to be' " [Seth, 2021a, 2021b].

The first thing in any attempted explanation of anything is to figure out and define what exactly there is to be explained. If we assumed that consciousness is everything, then we would have to explain everything. That would not be very productive any time soon.

Inadequate explanations result from inadequate definitions. Consensus definitions of consciousness are examples of those, they do not do anything good in the search for the explanation of consciousness.

The mystery of consciousness follows from vague definitions and the apparently immaterial appearance of the mind's overt content. Consciousness may be produced by the material brain, or it may be an immaterial phenomenon, which requires the brain in order to become observable. The current materialistic view is that the conscious experience is produced by the brain. But how would the brain actually do it; this has been the mystery.

A key concept in many explanations of consciousness has been emergence. When certain specific conditions are met or a certain system complexity is achieved, consciousness will automatically emerge. At first sight this may appear as a plausible explanation, but then it raises a question: How does this wonderful emergence happen? The claim that it just happens, does not explain anything at all. In fact, resorting to emergence as an explanation is an admission of ignorance.[1]

I think I am conscious — and I am right. If I were unconscious, I could not report that I am unconscious, I could not have any reportable thoughts or observations. When I feel pain, I am very much conscious of the pain. And, when I am awake, I am conscious of what happens around me.

The word "conscious" relates to situations like the above, and the word "consciousness" refers to the state of being conscious.

Note the difference between the experience of consciousness and the contents of consciousness. There is no need to involve the contents of consciousness in the explanation of the experience of consciousness. Also, cognition, like thinking, reasoning and imagination can be explained without the concept of consciousness. If the distinction between these and the experience of consciousness is not made, the end result will not be an explanation of consciousness, it will be a conceptual mess. Examples can be found.

Surely there must be somebody behind my eyes (see Chapter 9). Intuitively it might appear that this entity would be the consciousness. It would be an agent with executive functions, and it would constitute the self or the mind. After all, what would be the evolutionary gain of consciousness if it did not do anything?

Nevertheless, there are a number of researchers, who present that consciousness does not have any function, see [Brian, 2014].

Would it be possible that consciousness were a mere appearance that does not do anything. After all, our thoughts arise subconsciously; we are not able to observe consciously the processes behind these. Our thoughts just pop out to be rejected or approved. Also all our decisions are made subconsciously; conscious appearances being only the way in which these become into the focus of attention.

The explanation of consciousness has been searched for within the categories of immaterial substances, material properties, agents and operators. No convincing explanations have been found within these categories. The search for consciousness inside these categories is futile, as consciousness

---

[1]The author has not utilized and does not utilize the concept of emergence as an explanation of anything.

does not reside inside these; category errors have been made. Yet, the explanation of consciousness should be next to obvious.

18.2.   The Explanation of Consciousness

Is consciousness a material or an immaterial entity? If you think so, you may be in for a surprise.

As the first step towards the explanation of consciousness, the differences between the experience of consciousness, the contents of consciousness and cognition must be noted. Consciousness is not the same as the easily explainable consciously perceived contents of the mind; this content can easily be listed and classified. Consciousness is not the same as cognition, either. Cognition can be explained without any reference to consciousness.

When we are conscious, we can report that we are conscious. We can perceive the outside world, our own body and its conditions and also our mental content. We can also report our percepts to ourselves and also to others. When we are not conscious we do neither perceive nor report anything. Consciousness is the state of being conscious.

The nature of the consciously perceived contents of mind reveals what consciousness is all about; these contents consist only of percepts and nothing more. Consciousness is, if anything, the perception of instantaneous mental content, which consists of sensory percepts of the world and body, the virtual percepts of thoughts, imaginations and memories. All these have the appearance of immaterial qualia. All these are also remembered at least for a short while, and they are self-reportable in the sense that they can evoke other mental content and further thoughts; the flow of consciously perceived percepts will sustain the continuous flow of thoughts.

What is to be perceived, can only become perceived via the perception processes. The mental products of the brain activity are not products of sensory perception, but they can become virtually perceived, if they are returned to the sensory cortex areas. This can be achieved by internal feedback loops.

This is why we hear our thoughts in the form of silent inner speech and see our imaginations as vaguely seen images. As a result, all the contents of consciousness have the form of sensory percepts, virtual or real. And all these percepts have the common appearance, the qualia. All these percepts can be remembered for a while and can be internally reported.

When sensory perception and virtual perception ceases, like in deep dreamless sleep or during anesthesia, the consciously perceived content of mind becomes zero; we are not conscious of anything, and there is nothing to be reported. Subconscious processes may remain and need their own explanations, but that is another story.

Considering all the above, the author's conclusion is:

> *Consciousness is neither matter nor any immaterial substance; it is not an entity at all. Consciousness is nothing more than what it appears to be; it is self-reportable perception. Being conscious is the state of having the reportable flow of percepts with qualia.*

Qualia constitute the inner appearance of the contents of consciousness, and the phenomenal experience of consciousness is the flow of qualia; consciousness is self-reportable perception with qualia.

The thinking process in itself is very much subconscious, as we are not directly aware how the words pop up into our consciousness until our inner speech makes them virtually perceived. This inner speech is heard in the same way as real heard speech, which is perceived as auditory qualia. The inner speech operates with words, and these are perceived as virtually heard sound patterns.

However, the foregoing does not solve the hard problem of consciousness; the impression of immateriality of the contents of consciousness. This is a tough question, but a clever reader may notice that the answer to that question has already been presented in this book; the solution to the immateriality problem of qualia is also the solution to the hard problem of consciousness.

Qualia are the form of all contents of consciousness used directly as such or as symbols. When the qualitative and immaterial appearances of qualia are explained, then also the immaterial appearance of consciousness is explained. This explanation is given earlier in this book in the Chapter 8. According to that explanation, the immaterial appearance of qualia is not a difficult problem at all; it just results from our inability to observe the neural activity behind our percepts as what it actually is, a material activity.

Self-reportable perception with qualia constitutes consciousness, and the perception of one's self-related mental content constitutes self-consciousness; self-consciousness is self-reportable perception of one's own body and self-image.

Consciousness is neither a material nor an immaterial entity, it is not even an entity of any kind; it is mere perception.

18.3. The Seat of Consciousness

According to the Cartesian Theater model of consciousness, a small homunculus sits on the seat of consciousness inside the brain and watches the products of the senses and imaginations.

It is obvious that there is no such homunculus inside the brain. But how about the seat of consciousness? If consciousness were a distinct entity, then it might have also a distinct location inside the brain. For instance, the eyes project imagery on the receptor neurons of the retina at the back of the eyeballs, and these neurons transmit their information to the visual cortex. This, in turn, appears to transmit the information into the deeper neural layers in the brain. At some point the received information becomes consciously perceived; the experience of seeing arises. The seat of consciousness would be the area or a spot in the brain where this experience takes place.

It has been assumed that the finding of the location of the seat of consciousness would eventually solve the problem of consciousness; just find the location and inspect carefully what happens there. That would provide the explanation of consciousness, maybe. This search should be easy. Just provide a distinct sensory stimulus for the generation of a conscious experience. Then just observe by probes or by brain imaging machines, which part of the brain is activated. That would be the location of the seat of consciousness. This kind of research has been done. The results have been inconclusive, as many parts of the brain will be activated ay the same time.

There is another approach available. The scull and the brain can be extensively damaged, yet the patient may retain consciousness; thus the damaged area is not the seat of consciousness, the seat is elsewhere.

However, certain brain damage leads to the permanent loss of consciousness, while the patient would still be alive. Just locate this area, and that would be the seat of consciousness. These brain areas have been identified. They reside in the middle of the brain, and are the thalamus and the claustrum. Permanent loss of consciousness will result, if these areas are damaged, even when the brain is otherwise intact. The connection between the claustrum and consciousness has been studied among others by Crick and Koch [2005].

Except for smell, all sensory signals go through thalamus. It appears that the thalamus and the claustrum are actually pathways, attention controllers and distributors. It is obvious that when they are damaged, the brain's perception process ceases, nothing will be perceived and consciousness vanishes. The thalamus and the claustrum however, would not be the seat of consciousness.

In the foregoing it is argued that consciousness is reportable perception of sensory information and mental content, and has the form of qualia. This is a process, not an entity. As not being an entity, it does not have a seat, even though the main locations of this process would be found in the sensory cortex areas.

In the light of all the foregoing it is should be apparent that consciousness does not call for any supernatural substances or entities, and therefore there should not be any fundamental obstacles for the artificial generation of consciousness. Conscious robots should be possible.

18.4.　The Consciousness of Sensory Modalities

Is consciousness a unity? According to some philosophers and researchers, humans have one and only one continuous and unified stream of consciousness [James, 1890; Baars, 1997]. Introspection might superficially support this claim. At each conscious moment we seem to have a unified conscious experience of the totality of our percepts and other mental content; how else could we coordinate our sensory percepts and motor actions? For instance, it would be difficult just to walk in a busy city without being able to integrate visual and auditory information.

There are a number of theories that try to explain this by presenting consciousness as the unified and integrated whole of separate neural processes for sensing and acting.

One well-known theory along these lines is the Integrated Information Theory (IIT) of neuroscientist Giulio Tononi [Tononi *et al.*, 1998; Tononi 2004, 2008]. According to IIT, information integration will produce consciousness, as soon as enough information becomes integrated.

However, Tononi's and others' information integration theories do not explain consciousness because no amount of information integration will produce any experiences of qualia, and without qualia there is no consciousness. The hard problem of consciousness remains unsolved.

Then there are theories that speculate that consciousness were produced and unified by electromagnetic fields, or that consciousness were in itself a ubiquitous cosmic field, which would provide the unification. — It should be obvious that these kinds of theories do not explain much if anything; instead of explaining they introduce further elements to be explained.

In this book consciousness is not presented as an entity, it is reportable perception of the external and internal. The perception processes of each sensory modality do not need any overall unification; they do not need help from other sensory modalities. One can well be conscious, when blinded, or not being able to hear, smell or taste. In fact, one can be conscious without being able to see and hear, as is demonstrated by the cases of deaf and blind persons like Helen Keller (1880–1968) and Haben Girma (1988–). One can be conscious as long as one can perceive something, no matter how little.

Each sensory modality handles only its own kind of sensory information, and its internal qualia are experienced only within the modality, not elsewhere; for example, the qualia of the visual sensory modality are not experienced within the auditory modality. A word like "red" evokes the quale of red in the visual faculty, not in the auditory faculty.

Consciousness is perception with qualia, and each sensory modality has the perception process with its own kind of qualia; therefore it can be said that each sensory modality has its own consciousness. This claim follows directly from the definition of consciousness given in this book.

Here, the question is not exactly about information integration or unification; it is about the associative links between the modalities. Percepts from the different sensory modalities will be associatively linked as required by the cognitive processes of the mind. However, this association process does not unify any experiences; in the process of associative linking, qualia are not transmitted from one modality to another.[2] The associative connection only evokes percepts that have been associated with the percepts from another modality.

Association is not a unifier of sensory percepts, but the percepts from different modalities do have a common factor, namely their simultaneity. The flows of the percepts from the different sensory modalities are not actually unified, they are synchronized, just like the soundtrack and image in movies.

However, inside the brain there are two sensory modality processes that can unify everything from all sensory modalities, and moreover, within

---

[2]Synesthesia is a condition in which sensory stimulation gets mixed; sounds can be seen as colors and so on. But also in these cases the modalities produce only their kinds of sensory experiences.

their own sensory modality only. These processes are the *inner speech* and *imagination*. Inner speech resides in the auditory modality, and imagination resides in the visual modality. Especially, inner speech is the great unifier that can present everything in symbolic form with the power of a natural language.

18.5. Consciousness During Sleep

Are we conscious, when we are having a dream?

We are conscious when we are awake and are able to perceive our thoughts and are aware of our surroundings. We are not conscious, when we are in deep sleep and unaware of everything. However, during sleep some brain processes are still active. This has been shown with examinations by EEG (electroencephalography) machines.

An EEG machine is an instrument for the recording of the brain's electrical activity. EEG machines use electrodes, which are placed on the scalp. EEG examination is non-invasive and painless.

EEG recordings have the graphic form of waves of different amplitudes and frequencies from near zero to around 100 Hz. EEG recordings do not reveal thoughts as such, they only indicate excitation levels of large groups of neurons.

During wakefulness EEG waves have the frequencies mainly within the range from 13 Hz to 100 Hz. During sleep EEG wave frequencies go down to few Hz, but their amplitude may be high. However, during dreams the EEG waves and frequencies are rather similar to those during wakefulness.

Sleep is not an on-off phenomenon. There are the states of light sleep, deep sleep and rapid eye movement sleep (REM). Then there appears to be an altered state of consciousness, namely the "daydreaming state" with lowered awareness and possible dreams.

During deep sleep one is not conscious; no thoughts, no awareness of the environment, no sense of time. In this book consciousness is seen as self-reportable perception. If this perception process is cut off, consciousness vanishes. During deep sleep sensory perception and virtual perception are cut off, and there is no inner speech or other consciously observed mental content. The "contents of consciousness" will be empty. Unconsciousness will follow.

According to the terms of the perception-response feedback model, sensory perception will be cut off during sleep when the input thresholds are raised. Also the flow of inner speech will stop, when the internal feedback loops are no longer working.

Dreams occur during the rapid eye movement sleep (REM). When a person is asleep and has dreams, the person's eyes may be moving as if the person were watching something. This movement can be observed even when the eyes are closed.

A dream consists of the perceived flow of sub-consciously generated imagery and sounds. This can only be possible, if the internal feedback loops become active and allow thus the virtual perception of mental content. The sensory modalities are cut off, and cannot provide any reality checks. Without these, the constancies of objects and locations in the dream are not sustained, and the dream will become incoherent and surreal.

Dreams are consciously experienced, and can be remembered and reported afterwards, if the sleeper awakes soon enough. In the dream, the dreamer is the self-conscious experiencer and the main actor. Dreams are virtual personal experiences.

Dreams are quickly forgotten, as the short-term memories are not sustained long enough for the formation of long-term memories. Night-mares are remembered more easily due to their emotional intensity.

18.6. On the Existence of Consciousness

The conclusion that clarifies it all.

Incredibly enough there have been philosophical claims that consciousness does not exist. Also, there have been claims that consciousness were an immaterial entity inside our head, or it were an entity that hovers in and around our head, being at the same time internal, external and immaterial. Or, consciousness might be completely external, perhaps in the form of a universal and eternal consciousness field filling up the whole universe. The brain would only be an interface between the body and this ubiquitous field of consciousness.

But then, introspecting one's mind as much as one may, one can only find consciously perceived content, but nothing that could be called consciousness. Consciousness does not have an appearance. Consciousness is not an agent either, as it does have any functions. So what is left?

Regardless of any wild speculations, the strong view of the general public and many popular science writers has been that while nobody can ever explain what consciousness is, there is one thing that is certainly certain; consciousness does exist, and there can be no denying of this.

Neither of these opinions is necessarily true. Consciousness can be explained, but after that, and *just because of that*, there may be a serious issue with its existence, as is explained in the following.

Does consciousness exist? As soon as there is a rational explanation for perception, qualia and immaterial appearances, all the pieces of the puzzle of consciousness are there. If these pieces dovetail with each other and with cognition and emotions, the pieces are valid and the essence of consciousness is solved — and consciousness is found to be what?

In this book consciousness is explained as self-reportable perception in the form of self-explanatory qualia. This process would generate all the conscious contents of the mind, and in doing so it would also generate the impression of having a mind; without any contents there would not be any mind.

The neural basis of qualia is explained here as modes of neural excitations. The inner experiences of these excitations are also explained. Also the immaterial appearances of qualia and mental contents are explained. The immaterial appearance follows from the inability to observe the material machinery behind these; thus immateriality is misperception and an illusion.

Now to the unexpected conclusion: Without the explanation of qualia the explanation of consciousness is not complete, and when qualia are explained, the explanation of consciousness is redundant. Thus, as soon as qualia are explained, the mystery of consciousness evaporates; there is nothing more to be explained. There is no special "consciousness", there is only sensory perception and virtual perception of the mind's content.

Thus: We are conscious because we are able to perceive. The process that produces perception with qualia is just a process, not an entity. There is no need to assume an additional "consciousness" entity or contributor.

Consciousness does not exist; it does not exist in the forms in which it is usually presented and tried to explain. There is no such inner entity, material or immaterial that would count as consciousness. In science, the explanations of non-existing things are useless, so is also the naming of these. If something does not exist, its descriptions and explanations are as good as fairy tales.

All this has implications to the design of conscious robots. Self-reportable perception process is rather easy to design, and the illusion of immateriality is even easier; nothing has to be done. What cannot be perceived as material, seems to be immaterial. A conscious robot would be able to introspect its thoughts, but not the material machinery behind these. Thus, the thoughts of a robot would appear without any material basis, and therefore they would appear as immaterial to the robot as soon as the concept of immateriality was taught to the robot.

The word "consciousness" comes from the Latin word "conscientia", where "con" means "with" and "scientia" means "knowledge". Being with the knowledge about whatsoever is to be awake and conscious.

The word "consciousness" is formed like the words "happiness" and "sadness". The post-suffix "-ness" abstracts the original meaning. The word "consciousness" is truly abstracted, but not in good ways.

Abstractions are usually vague, difficult to explain and are prone to different interpretations. Abstractions are names, and naming may easily produce the belief that an actual entity existed behind the name.

"Consciousness" is a contaminated concept. It is contaminated by millennia of popular and philosophical misconceptions. The word "consciousness" has been used to refer to many different views and beliefs, which are not necessarily compatible with each other to any degree. This situation has secured the continuing philosophical debate around this phenomenon, as philosophers and laymen alike have their own individual concepts of consciousness. Consequently, any presented ideas about consciousness rarely match the preconceptions of others, no matter how accurate or inaccurate these were. The mismatch between the presented and pre-existing concepts may generate so-called mismatch displeasure and hostility towards the presenter of the deviating opinion. This situation applies to the concept of qualia, too.

Unfortunately, the term "consciousness" cannot be readily rejected, as it is so extensively used. A totally new unambiguous terminology for scientific studies of conscious states would be required. This kind of terminology is not yet available, if ever. Therefore the author is forced to use the words "consciousness" and "qualia" in this book.

Chapter 19

Natural Languages

19.1. Natural Language

Natural languages are spoken and written symbol systems that humans use for communication and information storage. Natural languages allow also verbal thinking and consciously perceived inner speech. Computer programming languages are not natural languages.

Basically, a natural language consists of vocabulary and syntax, the rules for the combining of words into sentences. Words are symbols, and as such they do not convey explicit meanings. This means that the mere memorization of the words does not suffice; the meanings of the words have to be learned, too. This is a nuisance, as every learner of a foreign language will soon notice. And even this is not enough; also the syntactic rules for the formation of sentences have to be learned.

Adults know that the learning of a foreign language is a hard task demanding lots of effort, time and the hardship of memorization. Yet, babies seem to learn their native language so effortlessly, and they will eventually speak fluently without any conscious idea of grammar and syntax. This follows from the different way of learning; babies learn the language through usage, not through rote memorization. Adults, in turn, will notice that the better they have learned to speak a foreign language, the less they think about the rules of syntax. The use of the syntax becomes eventually subconscious.

Languages may have very large vocabularies. The estimated size of the vocabulary of the English language may be around and up to million words, even though only a small fraction of that is normally used, or is even known to a common citizen.

A million words is a large amount, but not excessive. Finnish language has little bit more that is; up to infinity. These words are easily understood, even when heard for the first time. How can that be?

The grammar of the Finnish language[1] is very developed and quite different from most other languages. One special feature of the Finnish language is the creation of new words (as required or just for fun) from other words by adding suffixes and prefixes, and in this way altering the meanings of the words. The generated new word has a different meaning, but will nevertheless be readily understood, as the general function of the suffixes and prefixes is commonly understood, and they operate subconsciously. No conscious effort for this is required. This is so easy that even Finnish babies can invent new words, and they do it every now and then.

19.2. Learning to Speak and Understand Heard Speech

Babies learn to speak by hearing and imitating the speech of adults. Babies have high-pitched voices, while adults have low-pitched voices. How then can babies learn to imitate adults' speech, as they cannot produce the lower pitches of adult voices? And how would adults be able to understand the high-pitched speech of babies? There is an explanation.

The pitch of the human voice is the fundamental frequency or the lowest frequency of the vocalization. Males have low voices with pitches around 85–180 Hz, while female voices are higher with pitches around 165–255 Hz.

Basically, spoken words are sound patterns that consist of a succession of short vocalizations with distinct audio frequency spectrums. These vocalization units are called phonemes. The pitch does not carry information about the voiced phonemes, the information is in the spectrum of harmonics.

Pure sine waves do not have overtones, they have only the fundamental frequency. Harmonic frequencies are overtones with frequencies that are integer multiples of the fundamental frequency. The fundamental frequency can be understood as the first harmonic, the frequency of the second harmonic would be twice the first harmonic, and so on.

---

[1] English writer and scholar J. R. R. Tolkien noted: "Discovering Finnish was like discovering a wine-cellar filled with bottles of amazing wine of a kind and flavor never tasted before. It quite intoxicated me".

The spectrum of human voice consists of the fundamental frequency and its harmonic frequencies up to around 10 kHz. The highest harmonics have very low amplitude, and do not contribute significantly to the hearing experience and phoneme recognition. Therefore the higher frequencies can be omitted in narrow bandwidth communication, like in the telephone.

It has been noted that the pitch of a speaker does not contribute to the recognition of spoken phonemes. Instead, the feature to be observed is the relative intensity of harmonics within certain frequency bands, also known as formants.

More exactly, formants are the resonant frequencies of the vocal tract. The vocalization of a vowel leads to the generation of a certain intensity pattern of formants, and this pattern is rather similar for each speaker, regardless of their fundamental pitch. This allows the recognition of vowels whether vocalized by females, males or kids. (Fuller recognition of spoken words involves also the recognition of consonants and rhythm.)

Babies learn to recognize phonemes by their formants and to speak by imitating these,[2] not by imitating pitches.

Babies learn the meaning of their first words associatively, when their parents show things to them and pronounce the name of the thing at the same time. Later on, they will learn further words by themselves. Little children may understand words and speech before they can speak.[3] The understanding of more complex spoken messages is then another issue, there the mere recognition of words does not suffice there.

19.3. Learning to Read Written Text

Written text; the frozen speech that the reader makes alive.

All natural languages are initially spoken languages. Spoken words fade away and orally told stories will be forgotten eventually. The solution to this problem was the use of written symbols. There have been a variety of symbols for this purpose, like the hieroglyphs of Egypt and the cuneiform syllables of the ancient Babylonians. Characters, not letters are still used in Written Chinese and Japanese.

---

[2]This is a motor imitation act. The mechanism is explained in the chapter "Mind–Body Interactions in Robots".

[3]The author has tested this by asking a not yet speaking baby to show items: "show the table, show the pillow", etc. The child was able to do this with a self-satisfied smile.

Alphabetic or phonetic languages use a symbol for each phoneme. In a perfect phonetic language there is one-to-one correspondence between the pronounced phoneme and its written counterpart, the alphabet or letter. If you learn to pronounce each alphabet, you will be able to pronounce every possible word of that language correctly.

It should be self-evident that every modern language would be phonetic, but rather unfortunately that is not so. For example, English and French languages are very far from it; not all written letters are pronounced, and those ones that are, are pronounced in different ways in different words. Spoken and written English and French are practically two different languages. This delays the learning of writing and reading of the language.

In contrast, written Finnish language is perfectly phonetic in practice. Each phoneme has one and only one corresponding letter, and each letter of a word is pronounced, and always in the same way.

Learning to read a phonetic language is easy; just learn to recognize the letters and their connection with the corresponding phonemes. See the letter, and hear it voiced. Imitate this, until you remember it. Thereafter, the reading of a word is simply the voicing the corresponding phoneme of each letter of a word in succession, and there you have it; one simple operation works for every letter and word, even for any unknown words. In contrast, this operation does not work for non-phonetic languages, as letters and phonemes do not have one-to-one correspondence between them.

An advanced reader is able to see the text and recognize the words immediately without any obvious effort and letter-by-letter reading. That should suffice for information acquisition, as read words are as good symbols as heard words, and they should be able to evoke the meanings equally well. Nevertheless, most people seem to hear the read text as silent inner speech, too. Hearing the read text as inner speech is called subvocalization. Reading aloud is also readily possible.

If in the mind of the reader the meanings of the words are properly grounded, then the text can be understood without inner speech or any mental imagery.

Advanced reading is seeing, hearing and imagining, sometimes also feeling emotionally. This requires the attention, cooperation and coordination of sensory and motor modalities as well as emotional responses. Therefore it should be no wonder that a reader may not be able to pay much attention to the environment. Situational awareness and the scope of consciousness will be limited.

19.4. The Inherent Problem of Language

It is commonly understood that linguistic expressions refer directly to the matters of the world. Words and sentences describe directly whatever is linguistically reported, and modern languages with their large vocabularies would seem to be able to describe everything in the world. That would be a magnificent thing — if it were true. In fact, languages have a major problem, namely the ambiguity of meaning.

The meanings of the words have to be learned. Objects can be named, so also properties, actions, relationships and so on; the possible range of meanings is large. And that creates the problem.

What is in a name? "Blue" is the name for a specific color, and that is it, or is it? There is a problem; there are many hues of blue. "I want this to be painted blue" does not go without any further specifications. The word "blue" refers to a group of blues, and in that sense its meaning is ambiguous.

I got the blues, says the paint seller and the soul singer. Unfortunately, it turns out that most words do not refer to one well-defined thing only, but to groups of possibilities. This goes to nouns, verbs, adjectives and other word classes as well. For the correct understanding, the correct and relevant meaning has to be chosen, but how could this be done? Context, background knowledge and mental frameworks help here.

The meaning of the sentences like "it is there" cannot be understood without context. The words "it" and "there" depend on the situation at hand, and without that context the sentence cannot be understood properly.

"I can read the text and I know all the words, but I cannot understand anything". That may sometimes be a personal observation of an unfortunate student. What is wrong with the textbook, then? Nothing. The problem is with the student, who does not have the necessary background knowledge for the understanding of the text.

Mental frameworks are personal collections of concepts, ideas and beliefs about definite topics. Having a framework may help in putting the text in proper context. But frameworks may be counteractive, and lead to the evocation of meanings that may be quite different from what has been intended. Understanding the text may be difficult, if the reader does not have a required framework of the intended meanings. In that case the message cannot be put in correct context; it will become misunderstood or will not be understood at all. Abstract words like "consciousness" are examples of words that people understand in many different, even contradictory ways depending on the people's mental frameworks for that subject.

The heard and read words and sentences seem to refer directly to the matters of the world; that is our impression. But in fact, the words do nothing like that. They can only evoke whatever is associated with these in the hearer's and reader's minds. These associations do not refer to the world either; they refer to the products of perception and imagination. Even the sensory percepts do not refer to the world directly; they are only neural responses to external sensory stimuli from the world. What are actually named, are only our impressions of the matters of the world. Linguistic expressions are only expressions, not the actual things behind what is communicated. Exact linguistic communication is impossible. That is the inherent problem of language.

Nevertheless, without language there would not be the phenomenon of inner speech.

Chapter 20

Conscious Inner Speech

20.1. Inner Speech as a Conscious Phenomenon

Before the birth of spoken language humans did not have inner speech.

It is commonly understood that the silent inner speech (or inner voice) is the definite factor that distinguishes humans from animals. Animals cannot have inner speech because they do not have overt speech with vocabularies and language to begin with, even though some dogs and parrots may learn the meaning of few words. Parrots are even able to imitate spoken words and may be able to use them meaningfully. But these limited abilities do not suffice to maintain of any silent verbal flow of thoughts, the inner speech. Thus, obviously only humans can have this ability, which has been seen to enable higher consciousness, whatever that would be.

As a phenomenon, inner speech is the continuous silent speech that most people hear in their heads when they are awake. It is recognized as a product of one's own mind because it is inside the head, not coming from the outside, and it is generated by the hearer self. Hearing foreign voices in one's head without external causes is related to mental disorder called schizophrenia. Most people do not hear foreign voices coming from the inside of their heads.

Inner speech is inside the head, but without any definite location. The locations of heard sounds have normally locations outside the head, with one exception; when monophonic sound is heard through headphones, the sound has a definite location in the middle of the head. This experience is quite different from the experience of the inner speech.

Like the overt spoken speech, inner speech is neither actual thinking nor consciousness. It is the way in which the results of subconsciously generated linguistic thoughts become consciously perceived and become the focus of attention and evaluation. The actual thinking is a subconscious process.

Yet, inner speech is consciously perceived. If you are not conscious; that is, you cannot perceive anything, you cannot perceive and sustain inner speech either. The presence of reportable inner speech shows that a person is conscious; an unconscious person does not have inner speech and cannot report it. However, a person may be conscious also without inner speech.

Inner speech is a part of the contents of consciously perceived contents of the mind. Inner speech stops when we fall asleep and become unconscious.

Inner speech is also virtually spoken silent speech, which could be readily spoken aloud. This has a consequence; inner speech provides commands to the muscles of the speech organ, which obviously cannot pronounce two words at the same time. Only one set of commands can be executable at each moment, and therefore there can only be one stream of speakable inner speech at a time. Simultaneous vague auditory background experiences, like earworms in the form of songs and pieces of music, are possible.

Inner speech is silent self-talk, a verbal report of thoughts. Inner speech is understood without error as the speaker is also the hearer; the inherent meaning problem of language does not normally apply to inner speech.

Inner speech has connections with self-image, self-awareness and the awareness of mental content [Morin and Everett, 1990; Steels, 2003].

20.2. Inner Speech as Unifier

Natural language inner speech has a wonderful property; it can cover and represent everything that is known or imagined in the word. And all this is packed within one sensory modality, the auditory modality.

The words and sentences of natural languages are symbolic representations of the matters of the world. Syntax allows the manipulation of these representations without any reference to the outside world. Chatbots are a proof of this. However, without any connections to outside world, linguistic representations remain just collections of symbols, syntactically connected though, but without any meanings. The meanings of the ramblings of a chatbox are only in the human hearer's mind; the chatbox itself does not understand anything. Meanings must be acquired, and that is the task for sensory modalities.

Basically, sensory modalities produce qualia, the phenomenal percepts of the sensed. In the mind, these percepts are taken directly as what they are. But without any interconnections between sensory and motor modalities these impressions would be without any practical use.

Neural interconnections between the auditory modality and all the other modalities allow the inner speech; the use of sound patterns as words, which in turn are symbols for objects, properties and actions and so on. Inside the mind this process creates a virtual modality, the symbolic modality of natural language. This modality resides inside the auditory modality; it is a modality inside another modality. In this way the auditory modality has acquired a new function in addition to its basic function, the basic sensing of sounds.

Inner speech is virtually heard speech, and it takes place as a process within the auditory modality. But inner speech is not about heard things only, it is about everything. It is also about real or imagined matters of all the other sensory modalities. Inner speech connects sensory percepts and virtual percepts, feelings and memories, it connects everything. Inner speech is the great unifier.

20.3. Learning to Produce Inner Speech

Inner speech is not an innate process. Little babies do not have it because they do not yet have natural language and vocabulary. Babies learn to understand single words first, and they learn to pronounce them via imitation. Next, they learn to produce simple sentences.

Then usually happens something that the parents cannot fail to notice; children begin to talk continuously, especially when they are alone. This talk is not for communication. It is just a flow of trivial comments, statements and reports about the actual situation. In psychology this phenomenon is known as the child's overt self-talk.

It is usually assumed that children's overt self-talk is beneficial or even necessary for their cognitive development. Most probably that is so, but there may be another, possibly unavoidable reason; overt self-talk may be the child's only way of hearing her/his own thoughts because the child does not yet have consciously observable silent inner speech, see Fig. 20.1.

The basic auditory perception chain consists of the ear, preprocess, perception point and an association network that corresponds to the auditory cortex.

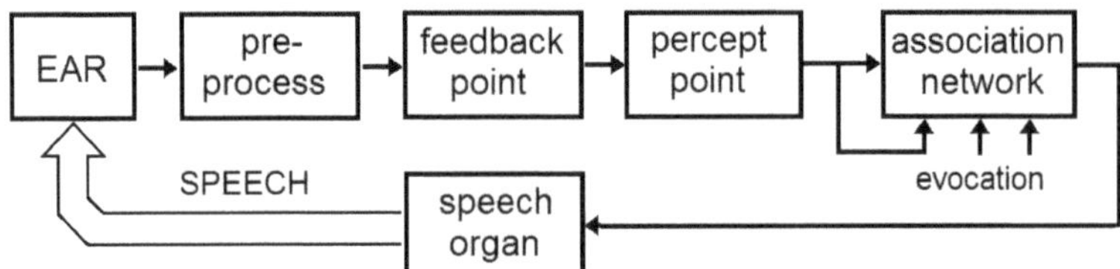

Fig. 20.1. The perception-response chain model. Hearing one's thoughts without inner feedback is only possible by talking aloud.

The heard sound is just an envelope of the sums of all heard frequencies, and as such rather useless. In the ear the cochlea produces neural signals for each distinct frequency with the resolution of about one hertz. This allows the identification of spoken words by their audio spectrum (and also by the rhythm).

The association network produces output responses, when it receives evocative cues from the percept point or from other sensory modalities. These responses are auditory neural signal patterns, and they are forwarded to the speech organ (and other sensory modalities). The speech organ takes these signals as motor commands, and produces corresponding sounds, such as words of the thought flow, the speech.

The produced speech will be heard by the system due to the external acoustic feedback from the speech organ to the ear. This is the only way in which this system can observe the products of the association network that is, its verbal thoughts.

Speaking aloud does enable the hearing of one's thoughts, but it is rather awkward way, as it also allows others to hear these thoughts. It would be better to keep one's mouth shut. This could happen, if the thoughts could be perceived internally, without speaking them aloud.

The child's overt self-talk ceases eventually. The child may have become aware of it, and may then feel kind of embarrassed. Or then, the overt self-talk ceases because the child does not need it anymore, as the child has become able to hear it without talking aloud. The child has developed silent inner speech. How did that happen?

Indeed, in the brain there appears to be a network that can enable consciously perceived silent inner speech; it is the very perception-response chain itself. What is required here, is an internal feedback from the association network output to the perception-response chain, see Fig. 20.2.

The proper feedback entry point would be between the output of the pre-process and the input of the perception point; after all, the output

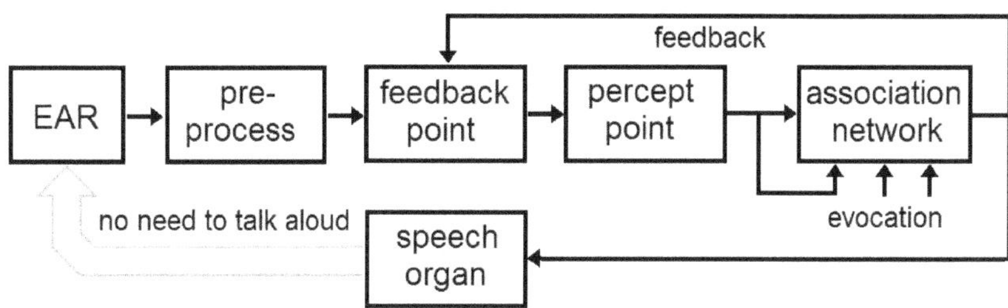

Fig. 20.2. The perception-response chain model with feedback. Internal feedback allows the hearing of one's thoughts without speaking aloud.

signals of the pre-process and the association network would be technically similar and representing sound patterns.

The feedback entry point would be a neuron group that would pass the neural signals from the pre-process and would also evoke these signals by the feedback signals.

However, there is a problem: There are very large number of neural connection lines between the pre-process output and the perception point input; how would the feedback signals know their target connection lines? There is no inherent way, if these connections are not innate, and in the brain, most probably they are not. These connections must be learned. A learning mechanism is presented in Chapter 23.

20.4. People Without Inner Speech

It is commonly assumed that everybody has a flow of continuous inner speech; how else could people know what they are thinking.

In reality the situation is not that simple, as the experience of inner speech is not the same for everybody. There are some people who have only limited inner speech, or no inner speech at all [Nedergaard and Lupyan, 2024; Makin, 2024; Hurlburt, 2011]. Then there is a group of people definitely without any inner speech; namely the congenitally deaf. The lack of inner speech is called anendophasia.

Congenitally deaf is a person who is born without the ability to hear. Due to their total deafness they do not have auditory sensory experiences. They have never heard people speaking, and have not tried to imitate sounds that they are not able to hear due to their disability. Thus they cannot extract sound patterns such as spoken words to be used as symbols. They cannot have verbal inner speech, and instead of that, they think using whatever working sensory modality percepts they have, such as imagery, gestures and sign languages.

Congenitally deaf people can learn to read printed text. To them printed words are just graphic symbol groups, which can evoke previously learned meanings in their minds. This learning may have taken place via ostension and association; words are associated with their meanings by showing the written word and pointing the intended object or action at the same time. Another way is to use a sign language and associate the signs with the written words. The ability to read text will allow verbal thinking, but not auditory inner speech.

Congenitally deaf people can produce sounds and can be taught to speak, but they will not hear their own speech and vocalizations, and will not acquire inner speech in that way.

Also normally hearing persons may have limited inner speech or no inner speech at all. It would appear that in these cases the feedback loop in the auditory modality is not working in the usual way; thoughts are not returned to the auditory modality perception point to be perceived and to become virtually heard speech in that way. Also, no runaway effect will follow. However, the thought signals may be transmitted to the speech organ and also to motor neurons, so that the person can communicate by speaking and writing.

20.5. Inner Speech as a System Defect

It might be inferred that since the inner speech does not specifically contribute to the processes of auditory perception, it would be a separate modality inside the auditory modality, which would then be the host.

This leads to a disturbing thought: Things that are inside a host and live from the energy of the host, are called parasites; is inner speech thus a parasitic process? And if so, what should we think about it?

As explained earlier, the flow of inner speech is facilitated by the flow of sensory percepts from the various sensory modalities and also by the associative words-words evocation within the auditory modality. Due to the internal feedback, the words and sentences of inner speech are able to and will evoke further words and sentences associatively in a serial way. This generates the sustained flow of inner speech. This happens also when actual sensory percepts are not present or are not attended to.

Internal feedback is useful in sensory perception-response loops in various sensory modalities, as it allows the augmentation of partial percepts and also prediction and anticipation; these functions are most useful for situational awareness. This applies also to the auditory modality, as it allows the recognition of partially heard words.

The focus and topic of the inner speech is determined by the actual situation and also by momentous needs, intentions and emotional states. Inner speech can be related to planning and problem solving, and will be sustained by these processes in themselves, as long as the attention stays focused on these.

Obviously, the flow of inner speech is useful for task-oriented thinking, problem solving, creative writing and some other things. But the question is: Do all these apparently beneficial actions necessitate the *persistent continuous flow* of inner speech? After all, there are apparently normal people without inner speech and without any apparent handicaps, which could be related to this condition.

Indeed, inner speech is not always task-oriented and serving some purpose. It can also be free-running, sustained only by the auditory modality feedback flow of associatively evoked words regardless of the overall meaningfulness.

Introspection seems to show that in the free-running inner speech, even irrelevant thoughts will evoke further thoughts and so on. This generates the runaway problem; the flow of free-running inner speech will become self-sustaining, and without much in the way of coherence.[1]

One may become deep in one's task-oriented thoughts, but that can also happen with free-running irrelevant thoughts. These serve no purpose at all, and as such they only will quickly fade away without any memory traces. That is useless waste of energy, and potentially harmful and even dangerous, especially when due to the runaway effect, attention cannot be paid to actually important or even urgent matters.

---

[1]The author's doctoral dissertation presented a simulation of a neural modular perception-response feedback loop cognitive architecture with various sensory modalities. It was able to handle simple text. The text handling "auditory" feedback loop was not normally closed, as the full closing caused an endless flow of incoherent verbal "inner speech" [Haikonen, 1999]. The same effect was noted during the construction of the author's XCR-1 robot. There the visual feedback loop is closed and free from runaway effects, while due to the runaway effect the auditory feedback loop is not closed.

It is quite possible that the incoherent runaway inner speech does not serve any practical purpose; it only consumes energy and disturbs attention. In the engineer's eyes, there might be a system design issue here. In feedback control systems, runaway effects follow from faulty designs; uncontrolled runaway effects are system defects. But sometimes system defects may also bring in unexpected benefits.

Chapter 21

Towards Conscious Machines

21.1. Conscious Robots

Mens conscia ex machina. Conscious mind from the machine.

The engineering wisdom has it that we cannot truly understand and explain a phenomenon until we have reproduced it artificially; the proof of the explanation is in the replication. This wisdom applies to conscious states, too. We have to construct conscious humanoid robots in order to truly understand what it takes.

Humanoid robots are robots that look like humans, and they are here already. They can be mainly found in robotics laboratories. Some of these robots can walk, run and execute some chores, some may just sit and look nice. They may talk and understand speech. They may also be able to mimic human facial expressions. Their cognitive abilities are powered by AI. These robots are impressive achievements in engineering enabled by high technology.

Applications of current style humanoid robots are seen in customer services, health care and possibly in companion, helper and aid services for the elderly and incapacitated. So far however, the limited practicality and high cost of these robots have prevented their wider use.

Yet, as advanced as the contemporary humanoid robots might seem to be, there is still one thing missing, and very essential thing it is. The self is missing; inside these robots there is nobody at home. These robots are not conscious, they do not know what they are doing, nor do they know that they exist. They do not have minds. They are only automata; advanced ones perhaps, but still mere automata.

Philosophical reasons aside; if these Artificial Intelligence robots really do or will do what they are supposed to do, then why should they have to be conscious? Does it really matter to the human master of the robot, if the robot is able to produce the plausible appearance of a conscious person on its own right, and is able to converse with its human master like any human? After all, the human master does not have any simple way to test what is going or not going on inside the robot's head. The robot might be or might not be truly conscious, while the external appearance and behavior would be the same. So, from the human master's point of view, what would be the practical difference, if any?

The practical reason is simple. To be able to work efficiently and safely in real world situations and environments, human-like robots should possess true situational awareness. But unfortunately, without being phenomenally conscious there would not be any such awareness. Mere external appearances are no good, when in an unfortunate situation the robot responds badly and creates damage and mayhem. Robots should be conscious for many reasons [Aleksander, 2009, 2015].

Examples exist: Self-driving cars based on Artificial Intelligence were supposed to eliminate all traffic accidents, as computers would not become tired and could stay alert all the time, while human drivers are known to lose their focus of attention every now and then. Nowadays AI self-driving cars exist, and so do accidents caused by them. Self-driving cars may remove human errors, but they are quite able to produce their own, as we have seen.

What then would it take to create conscious machines and robots with minds? What would be the crucial problem to be solved first?

The crucial problem is not about the robot's cognitive abilities. It is not about a chatbot's ability to put words after another, either. Instead of all this, it is about the self inside and the mind with its inner appearances. It is about the feel of existing and "living" in each situation; a conscious robot must have the experience of being the one behind its eyes. Would all this be artificially achievable by computational methods and computer programs, or would non-computational methods be necessary?

21.2.   Computational and Non-Computational Approaches

Could a computer program be conscious?

One of the tenets of Artificial Intelligence has been the Church–Turing thesis, which has been taken to prove that any computation executed by a computer can be executed by any other computer, at least in principle [Church, 1936]. Consequently, if the brain were seen as a computer, then also the brain's workings would have to be computable by any other suitable computer. Therefore consciousness is computational and can be programmed; the computer can become conscious.

Actually, the Church–Turing thesis neither said nor logically implied that this would be so, it just effectively stated: Anything that a computer can compute using general recursive functions is also computable by another computer, which also uses general recursive functions, even when the computers were otherwise different. Church–Turing thesis is about computation and the requirement of memories for the storage of intermediate results for their later access when they are needed in the course of the calculation.

It should be self-evident that Church–Turing thesis does not relate to the problem of conscious machines in any way, but perhaps consciousness might nevertheless be computational and achievable by computations for other reasons.

Some contemporary researchers argue that consciousness is necessarily computational [Baars and Franklin, 2009], and there is no other way to create it artificially, "computationalism is still the only game in town" [Davenport, 2012].

Thus, according to computational theories, consciousness will arise whenever a physical system, like a digital computer, executes a *suitable computation*. And here, "a suitable computation" is a computation that is able to produce consciousness.

Of course, all this depends on how "consciousness" is defined. If consciousness is defined as a *computational process*, then the claim that consciousness is computational, is a tautology and true by definition. This conclusion has had some appeal to AI programmers. But unfortunately, this is circular reasoning and a logical fallacy.

Computational processes are computational by definition, but this does not say anything about consciousness. In reality, conscious experiences are not computational, and this fact should be obvious to anyone who has hit her/his thumb by a hammer.

But, if you cannot make it, fake it. If you want to fake it, simulate it. Computational theories of consciousness assume that consciousness is a physical phenomenon, and as such obeys the laws of physics. The laws of physics are computational, and allow the computational simulation of physical processes. As this is so, then also artificial consciousness can be produced via the simulation of real consciousness.

Everything can be simulated with computers. If weather can be computationally simulated, then why not also consciousness? Weather forecasts are nowadays based on computer simulations, which are nice and sometimes even almost accurate.

Surely we could build a simulation-based robot with all the external signs of being conscious, and for an outside observer that should suffice.

Computer simulations are just mathematical models. A model is just a model, not the real thing. For example, the strength of pain can be mathematically modeled and can be given a numerical value, but that is not the same as the experience of real pain. Simulated pain does not hurt, especially if there is no conscious experiencer.

I can do some computations, but I am not conscious because of that. I am conscious because I am the one who is behind my eyes. I would like to see the piece of computational code that would produce these kinds of *phenomenal experiences* inside the computer's mind. No such code has been found and invented, as inside the computer there is no mind with phenomenally felt experiences to begin with.

Of course a computer may be programmed to "announce" that it has the feeling of being an entity behind its eyes, or something similar, but that is beside the point. The computer still does only what the program commands, and it does it blindly, without any awareness of what it all is about. The computer does not have what it takes to be a conscious entity, and no fancy cosmetics will make any difference here.

The counterargument here would be: It would not be the computer's hardware that were conscious, it would be the program that were the conscious entity. Thus, so very conveniently, this artificial conscious entity would also be kind of immaterial, as opposed to the material executing hardware. This sounds alluring, but unfortunately a computer program is just a list of commands, which make the computer's machinery to execute the commanded functions. These functions are executed, but not mentally experienced by anything inside the computer; computer programs are not experiencing entities in any phenomenal sense. And computer programs that simulate the workings on the brain are not any experiencing entities either.

Being conscious is more than mental computation and arithmetic operations, and Church–Turing thesis has nothing to do with the self, the mind and phenomenal experiences. And these are the very aspects that cannot be overlooked or omitted in the development and artificial reproduction of consciousness. Computer simulations of the workings of the brain do not lead to the desired outcome, either; they will not produce any inner phenomenal experiences in the computer. Being conscious is not computational, and it is way beyond the capabilities of computers.

What if there were another information processing method, one that would not be based on programs and computations? That method should solve many problems of machine cognition, and might also allow the creation of artificial entities with the true experience of "the self behind their eyes". That should not be impossible, as there is already an example of this kind of material system, which does not operate with programs and computations, namely the human brain with its immense number of neurons.

Since the early days of AI, the modeling of the brain with artificial neural networks has been seen an alternative way for information processing. Various layered neural networks and related synaptic value adjustment algorithms, such as the Back Propagation and Deep Learning, have been designed mainly for pattern recognition and classification purposes. Also neural associative memory networks have been designed, such as content-addressable memories, linear associators, Hopfield networks, Kohonen associative memories.

Usually though, the original idea of non-program distributed systems has been forgotten, and these neural networks are realized as computer program simulations with complicated algorithms. In this way the originally seen benefits of parallel neural processing and the operation without computer programs have been lost and forgotten. But that does not seem to matter, as the missing of the lost benefits has been compensated by the unforeseen computation speeds of modern computers and the parallel use of many processors.

Thus, these above approaches are being successfully used for well-defined individual tasks, as the generously hyped generative chatbots show. However, they will not constitute true universal cognitive systems; they and their combinations will not emulate the workings of the whole brain, and even less the operation of the mind, let alone achieving consciousness and putting someone behind the eyes.

The original vision of non-program neural cognition has not been realized, and there is a reason for that. The design of conscious machines is not about the design of single task "neural" pattern learners and recognizers and their computational realizations. It is about the design of whole cognitive systems with minds and the capacity to have the experience of the conscious self.

More specifically, these kinds of systems would not be any computer simulations of neural networks. They would be real assemblies of non-computational associative neural networks based on dedicated hardware neurons and circuits operating without any program code. They would be able to put someone behind the eyes of a robot.

As explained earlier in this book, being phenomenally conscious involves the existence of the subjective experience of self and sensory perception with self-explanatory and self-reportable qualia. Pure symbol processing systems cannot provide these. They can only produce external make-believe appearances of mind and possible compassion for an ignorant and gullible external observer. There are also other related goodies that the current computational AI cannot deliver. Therefore the computational AI approach toward conscious machines has to be rejected; it will not be able to deliver true conscious machines. Instead of numbers and program codes, a conscious machine should work with inner appearances.

What is this pursued *machine consciousness* then? It would be much clearer, if instead of the word "consciousness" the expressions like "being conscious" were used. "Consciousness" is an abstract concept without any definite meaning, and it has the ability to create confusion.

21.3.   Phenomenally Self-Conscious Robots

I am me, and I am a robot.

A human-like conscious robot will have to have the "I am me" or the "I am the one behind my eyes" experiences. The "I am me" experience is an expression of self-consciousness, which property is understandably quite necessary for functional humans and also for conscious robots.

One cannot be self-conscious, if one is not conscious. The state of being conscious is to have reportable perception, and this state enables also the perception of matters related to the self. As explained earlier, the self is not actually any entity that observes, it not an entity at all. It is an impression and subjective experience produced by perception.

Human-like conscious robots with the internal impression of self should obviously be modeled after human beings, as the desired mental experiences should be similar to those of humans, or even better.

The human experience of self involves the sensory experience of the external world and the experience of one's own body and mind. These experiences arise from the sensory perception of the world and the virtual perception of the mind's content. The distinction between these two allows the concepts of the external world and the self. The experience would be: I am inside my body, and my body is separate from the environment. My mind is my mind, and I see myself as the one who controls and commands my body and its actions in the environment.

A conscious robot would utilize various sensors for the detection of its body part positions, material inertia, weight and the limits of the physical effectors. These percepts would be in analog form.

The sensory percepts of the robot body and its stress would also be unique; they would have a feel, which would be pleasant or unpleasant.

The sensory percepts of the surroundings would be different. These percepts would not only about the external; they in themselves would be externalized into impressions of the things out there.

A conscious robot would notice the difference between the surroundings and the robot itself. When the robot moves around, the environment stays put, it does not follow the robot. The robot would perceive its body as an invariant, which follows wherever the robot goes.

Just like humans, a sensing robot is a moving vantage point in the middle of the environment. Robot's visual perception would allow the first person point of view over this environment, and in doing so, it would also create the impression of the self behind the eyes. This first person point of view may also trigger the "I am me" experience.

The concept of self will arise from these kinds of observations, and will be amplified by the power of natural language and inner speech, the silent self-talk, by the association of the general words "self" "I", "me" and so on for the self-related matters.

Later on, the memorized personal history will contribute to the experience of self and self-image.

A conscious robot should have similar experiences, and these experiences should generate the concept of self for the robot. The first person point of view would be very important here.

Being self-conscious is a by-product of being conscious; you cannot be self-conscious, if you are not conscious.

These issues should be taken into consideration in the design of true conscious robots and system architectures for conscious robot brains.

Chapter 22

Haikonen Cognitive Architecture: An Associative Architecture for Robot Brains

22.1. Cognitive Architectures

The proof is in the design.

A system architecture is a formal representation of the interconnections between the different modules of the system. System architectures may be metaphorical, theoretical or physically implementable. Metaphorical system architectures have been seen in philosophy, but in practical engineering metaphorical architectures are useless.

The issue of implementability hinges on one aspect: Is it possible to realize the various modules of the architecture by using actually available hardware components or not? If not, then the system architecture is not implementable at least for the time being, and as such it would only be a theoretical or hypothetical concept with limited practical value.

System architectures are often represented in the form of block diagrams with interconnections, so that the overall operation of the system can be visually determined. The blocks are usually depicted as small boxes or squares. The blocks are named, but their internal workings are not depicted. This is done separately, and in electronic systems these depictions may be in the form of circuit diagrams, which are implementable.

Cognitive architectures are models and descriptions of systems that are aimed to produce cognition and possibly also consciousness. Cognitive architectures have been used in the research of machine cognition and consciousness. In the recent years, several more or less detailed cognitive architectures have been presented, e.g. the Global Workspace model of

Baars [1997], the Shanahan architecture [2010], Sanz *et al.* model [2009] and the Kinouchi and Mackin architecture [2018]. A comprehensive list of cognitive architectures has been presented by Samsonovich [2010]. All these have their shortcomings, especially what comes to the issues presented earlier in this book. I it were not so, we would already have true cognitive and conscious robots.

The author's HCA architecture is one example of architectures that are designed towards the creation of cognitive and conscious robot brains.

22.2.   The HCA Architecture

The author's Haikonen Cognitive Architecture (HCA) is an associative architecture for robot brains, see [Haikonen, 2007, 2019]. It is a neural system for the production of sensory perception, humanlike cognition with inner imagery and inner speech. The principles of the HCA should ultimately allow the creation of conscious robots in the terms of the principles presented in this book.

An early version of the HCA was presented in the author's doctoral dissertation [Haikonen, 1999] and has been elaborated in the following years. The basic principles of the HCA have been successfully tested in the author's XCR-1 hardware robot [Haikonen, 2019].

The HCA is a modular architecture that combines sensory perception, cognition, emotions, memory functions and motor action control into an integrated system. The HCA has perception-response feedback loop modules for each sensory and motor modalities. These modules are associatively cross-connected in the style described in the Chapter 15.

All these modules operate associatively with dedicated hardware neurons and neural networks. There are no code, no programs, no microprocessors, or microcontrollers.

The HCA operates with sensory percepts. These are non-symbolic and self-explanatory, and the system will take them as properties of the sensed. These percepts can also become used as symbols, and in this way the seamless transition from sub-symbolic to symbolic processing exists inherently in the HCA.

A simplified block diagram of the HCA architecture is depicted in Fig. 22.1.

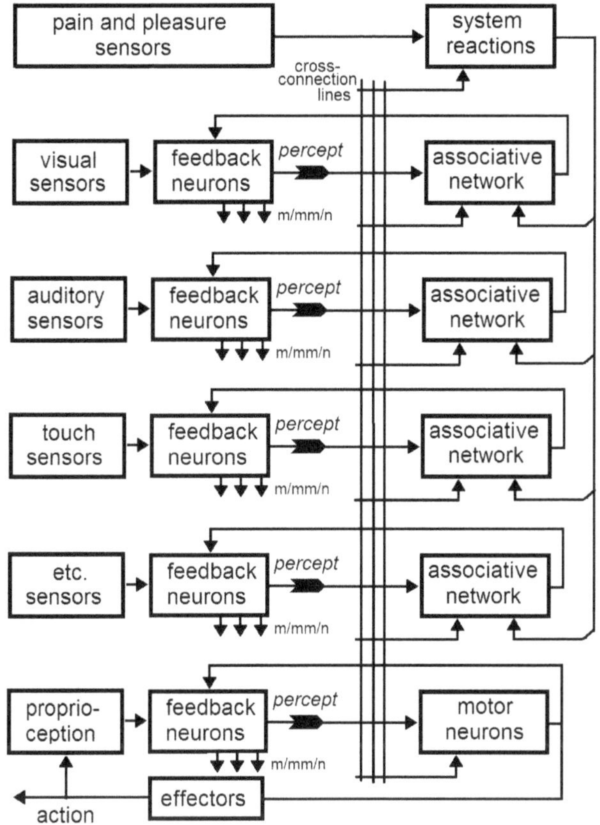

Fig. 22.1. Simplified overall block diagram of the HCA.

The HCA consists of associatively cross-connected perception-response feedback loop modules. Each module consists of a large number of parallel single signal feedback loops.

The HCA includes also pain and pleasure experiencing system, which transmits signals to the sensory modalities, allowing the association of percepts with pain and pleasure and also the emotional control of threshold values.

Each percept consists of a pattern of feature signals. Interconnecting lines depict paths for signal patterns. Apart from sensors and effectors the blocks depict associative neuron groups.

Associative networks are able to learn and generate expectations for what is being sensed. These expectations are transmitted via the feedback lines and are compared with what is actually perceived. Corresponding match, mismatch and novelty conditions (m/mm/n) are detected and these are used to attention control. These conditions are reportable.

The HCA is not a theater model. There is no common "arena" or "theater stage" for the sensory percepts to be observed by any overseeing entity or assembly. Each sensory modality of the HCA is independent and senses and experiences only its own kind of phenomena; these are not shared as phenomena with the other modalities.

The interconnections between the modules are only associative links, and as such they do not deliver anything to the receiving modules. Information is not shared, it is only cross-connected. As a result, the disconnection of one or some modules does not affect the remaining modules; they are able to continue functioning normally. The overall performance of the system will suffer, but the system will not crash. The human brain works in a similar way, and this can be demonstrated: The closing of eyes will disable seeing, but all the other senses will work as usual. The same applies to hearing and other senses, too.

Each module of the HCA operates with associative neuron groups, which use the Haikonen associative neurons. The Haikonen neuron is able to associate one signal with another or a pattern of signals with one signal. These neurons learn and operate individually. The synaptic values of neurons are not tweaked against each other, and there are no synaptic value tweaking algorithms either.

Each sensory modality has its own large associative neural network. This network receives input only from inside the modality itself; thus the handled information is always in the terms of the modality's special topic. Visual modality handles information only in visual terms, auditory modality handles information only in auditory terms and so on.

The percept points in every modality are the only places where the entities (sensory percepts, inner speech, imagery) appear as sensory experiences. These experiences are not transmitted to the modality's associative neural network as such, only neural patterns are.

The associative network does not operate with explicit meanings (which would be the experiences), it is only a vast connection network, where neural signals, signal patterns and sequences are associatively connected and evoked. There are no memory cells or locations, or stored memories as such. Perceived memories are generated by associative reconstruction at the feedback points.

Waves of neural activations may criss-cross the network, but inside the network there is no unit or entity that would know what this is all about. The actual meanings of the outputs of the associative networks are restored via feedback to the perception points.

Associative linking and evocation alone are not sufficient for cognition. Additional functions like match, mismatch and novelty detection are required for reasoning operations, and are implemented in the HCA in neural ways.

In large associative systems almost everything may eventually be linked with everything else, therefore there has to be means to keep the focus on the current topic, and also means to change the focus of attention quickly, when the situation so demands. In principle, the strongest signal patterns are attended to, but also new unexpected signal patterns may capture the focus of attention.

Percepts may have learned emotional significance; things may be good or bad. This significance is one important attention control factor.

Innate and acquired needs are motivations behind actions, and evoke related associations and actions.

It is useful to note here the difference between the associative neural networks of the HCA and the traditional artificial neural networks. The latter have used artificial neurons like the McCulloch–Pitts neuron [McCulloch and Pitts, 1943] and the Rosenblatt Perceptron neuron [Rosenblatt, 1958]. These neurons can be used in pattern recognition circuits with computational synaptic strength tweaking algorithms such as the Back Propagation and Deep Learning. These neurons and net-works are not suitable for true associative processing. Due to these differences, the operation of the HCA cannot be understood in the terms of traditional artificial neural networks.

More detailed description of the HCA and technical details as well as practical realizations of the author's artificial associative neurons and associative neuron networks can be found in [Haikonen, 2019].

22.3.　Conscious Immaterial Mind in the HCA

Mind is not a distinct entity inside the brain; it is an abstraction. Mind cannot be observed as such, only its contents and so-called mental states can be. The consciously perceived contents of the mind consist of sensory percepts and virtual percepts as well as the states that are related to emotions and moods.

The first "I am me" experience is a decisive moment in the arising realization of one's existence as an individual, a self. The subjective experiences of me, self and mind arise from perception processes and the insider-outsider difference. This difference arises from the internalization and externalization of the location of percepts. "Me" and "self" relate to the ownership of the body and the experience of being an individual, and the one who is behind the eyes (see Chapter 9) and the one, who experiences. "Mind" is related to the experience of having mental content and opinions.

Any allegedly conscious cognitive architecture would have to be able to cover these issues. That applies to the HCA as well. But, architectures are not conscious, their physical implementations may be.

Implemented architectures for self-awareness have to be embodied; without a body the impression of having an own body and self will not easily arise. The impression and effects of the insider–outsider difference cannot arise either. A conscious robot has to have a perceivable body.

It would be good, if the robot could see its body, but the mere seeing would not be sufficient for the impression of the ownership of the body and its parts. Touch and pain sensory modalities and motor modalities would have an important role in this. A robot needs touch and pain sensitive "skin" and hands with touch sensitive fingers for the inspection of its body. Touching external objects will deliver sensory experiences of that object, but touching oneself will deliver something more; namely two sensations, one at the touching fingertips and another at the touched part. Too hard touching causes pain. This definitely leads to the concept of self; if it hurts, it is me. This contributes also to the internal–external division; it does not hurt when something bad happens to external objects, but it does hurt when the same happens to me.

In this book it is presented that consciousness is reportable perception. In the HCA the perception process is designed to produce this kind of function. HCA is also able to support inner virtually heard speech and virtually seen imaginations.

In the HCA architecture, "mind" would be apparently immaterial, as the material neural machinery would not be observable by the system.

22.4.   The HCA and the Human Brain

The HCA is a biologically inspired model and architecture. However, it is not a model for the biological human brain; it is a model for non-digital robot brains with artificial neural networks of associative hardware neurons.

Brain research has been able to detect distinct areas of the brain for some distinct operations. There are sensory cortex areas for vision, auditory perception and other sensory modalities. The frontal cortex has been seen to process meanings and is assumed to execute higher level executive tasks, such as problem solving, planning, organizing, cognitive reasoning, decision making, and goal attainment [Miller and Cohen, 2001].

The concept of information integration is often mentioned in this context, but otherwise, it seems that current brain research does not yet understand how the neurons and their networks execute these tasks.

It has been experimentally shown that the HCA does work at least at the level that is implemented in the XCR-1 robot. It is also known exactly how the HCA neural networks do this. Therefore it might be asked, if HCA would offer any insights to the brain research.

HCA has separate sensory modalities with feedback loops. Feedback loops can be found also in the brain, possibly with similar functions.

Thus, in the HCA architecture the associative network areas with their cross-connections would more or less appear to correspond to the frontal cortex, when only inspected by the brain's neural wiring.

But there is the thing. The brain's frontal cortex is seen to process meanings, while the associative networks of the HCA do not handle meanings at all; they only learn and evoke connections between signal groups and modalities without any knowledge about what these signal groups might represent. These associative networks are only connection networks, while the actual meanings can be found only at the sensory percept points. The percept points deliver signal patterns to the associative networks, which will then send back their responses via the internal feedback lines to the percept points. There the responses are now returned into virtual percepts and are evaluated in the same way as the real sensory percepts. The match, mismatch and novelty conditions are detected and also emotional significance is taken into account. All these contribute also to the attention focus control.

The associative networks in the HCA do not think, they are only manipulators, which facilitate the associative search and evocation by cross-modal connections required for cognition. All this is subconscious. This process is fast because it is parallel, and it is efficient, as it utilizes methods for the selection of the most suitable output to be fed back to the perception process, where the results of this process get their meaning by appearing as virtual percepts.

The point is; what if the brain's frontal cortex operated in a similar way, not being a meanings processor, but just a vastly parallel connection network, a sidekick to the sensory modalities? In that case, meanings and qualia would reside only in the sensory cortex areas. That would make sense also from the evolutionary point of view.

The following may amplify the foresaid.

Chapter 23

Mind–Body Interactions and Self-Wiring in HCA

23.1. Mind Commanding the Body

Nature provided us with hands to interact with the physical world.

Without working muscles we would be dead; breathing is an interaction with the physical world. Work is another.

It is obvious that the hard work we do with our hands is executed by muscles. But also spoken words are produced by the muscles in the speech organ. Handwritten text on paper is produced by the muscles of fingers. Painters' artworks are produced by muscles. Our legs allow us to move around. When it comes down to it, muscles are ultimately our only natural means to execute interactions with the world.

We can plan our physical actions because we have brains that are able imagine. Technically, the brain is connected to our muscles by a large number of neural fibers, which carry motion commands for the execution of the intended actions. There are also other neural fibers, which carry information about the muscle tensions back to the brain. These connections work well, and the actions and motions are executed effortlessly. But behind this effortlessness there is an enigma.

It is known that the activation of certain neurons in the brain leads to the excitation of certain muscles and the production of motions. Thus, when the actual execution of an imagined motion is desired, certain neurons should be activated. But how could this be done, and which neurons should be excited? After all, we are not aware of our own neurons, and are not able to excite these directly at will. The actual neural operation is subconscious and beyond our conscious command and observation. We are not aware of it and do not consciously plan it. We only see the possibility of an action, and without any further ado the action can become executed.

One may propose that there is no problem here; neuroscience shows that muscles receive proper commands from the brain via neural fibers, and that is it. But, in reality that is not so simple at all. Behind the apparent easiness of muscle control hides a deep fundamental issue about the workings of the brain, and this issue is related to the mind–body problem; how the mind controls the body. This issue cannot be avoided in conscious robots, either.

For example, consciously perceived imaginations take place in the visual cortex area in the brain, while motor actions are controlled in a different area, in the motor cortex. Visual imagination signals do not control the muscles; this is done by neural signals from the motor cortex. Therefore visually imagined actions have to be translated into the corresponding motor commands, but how?

The brain is a vast collection of neurons. These neurons know nothing. In the brain each neuron is an individual unit, which receives stimuli from other neurons via the axons and synapses and delivers output signals through its own axon fiber. The neuron just is there and reacts, but it does not know where the input signals come from, what is their meaning and where its output signal will go to.

There are many, many fibers that connect the brain with muscles, but none of these fibers are labeled. The fibers are fixed, but the neural network of the brain does not know to where each fiber goes, and whether its endpoint is connected with a receiving party or anything. The receiving muscles do not know either, where the fibers come from. In fact, they do not know anything about the neural fibers. They only react to the impact of the received signal.

Secondly, the imagined vision of the desired motion does not include any consciously or subconsciously perceived knowledge about neurons and fibers, let alone knowledge about which neurons and fibers should be activated.

This situation applies also to the individual neurons involved in the imagination; they do not have any knowledge about the relevant neural fibers, either. This leads to an obvious problem; how to connect the brain's commanding neurons with correct muscles, when the system does not know what is to be routed to where and which connections should be activated.

Thus, the mind–body interaction between imagined actions and the muscles would appear to be next to impossible. Yet, it is not, as everyday practice shows. The mind commands and controls the muscles effortlessly and without any explicit or implicit, conscious or subconscious knowledge of the executing neurons and their connections. This kind of subconscious control of muscles can and must be learned early in one's life. This can happen, as the brain of a newborn has a vast number of neurons and synapses waiting to be connected; in the course of early development the baby's brain will self-wire itself.

Following examples show how this is done in the HCA, and might be done in the brain.

23.2. Learning to Reach Out

Mapping the space around and reaching out for something.

There is a cup on the table, in front of an AI robot. The robot is commanded to pick up the cup with its mechanical hand. The robot scans the area with a laser scanner and computes the numeric coordinates for the cup using a preprogrammed algorithm. Then it computes the required command values for each of the involved motors using another pre-programmed algorithm. Feedback control algorithms may be used to make the robot hand to home in on the cup. When the computations are completed, the cup is picked up. Microprocessors are very fast, and the robot can execute this task and others very quickly.

We humans may also want reach out and pick up cups and things. However, the neurons in the brain are rather slow, and the required computations would take some time, too much in many cases. This is evident to anyone who has done mental arithmetic. Besides, the brain is not a computer, and does not inherently do any numerical calculations. Yet we are able to execute these tasks quickly; just see what is there to be done, and it is done. We do this without any innate algorithms and programs and also without the computation of numeric coordinates for anything. Wherefrom could we have acquired the numeric values and learned to do these calculations in the first place?

There is an alternative way for computational control of muscles; the human non-computational neural way, and it is fast — and unobservable directly. This way is learned early in the childhood.

Muscles can do two things; they can contract and expand or relax. These two effects can be caused by involuntary reasons, like cramps. The contraction and relaxation of muscles are also controlled by neural signals from the brain. The degree of muscle contraction is determined by the intensity of the stimulus. When the stimulus ceases, relaxation follows.

For instance, the angle between the forearm and the upper arm is determined by the contraction degree of the arm's muscles. The contraction is, in turn, controlled and sustained by the excitation pattern of motor neurons. An associative connection between the seen locations of things and the corresponding hand motor neuron excitation patterns would allow the grabbing of nearby objects.

Infants execute haphazard and apparently meaningless hand motions, but after a good while they will be able to produce coordinated motions, such as the reaching out towards a given position and the grabbing of an object; the haphazard hand motions have amounted to the mapping of the seen reachable space around the infant and the association of the random motor commands with the reached position.

In the act of reaching out, the thing to be reached is not the actual target; the actual target is its position. Therefore, if the position of the thing is known, it can be successfully reached out also in total darkness or eyes shut, and even using different hand trajectories, as the trying of this will show.

This assumed learning process with random motions can be readily incorporated in HCA based robots. There the random motions are caused by random motor command excitations, which will lead to random hand positions. These are seen and perceived. Visual perception does not only produce image percepts, it also produces the impressions of positions; what is where in relation to the body.

In the HCA based robots the visual position percept signals are associated with the simultaneous hand actuator contraction signals. In this way, haphazard hand motions will eventually lead to the mapping of the reachable space in terms of hand actuator, and the random excitations will fade away.

After the completion of the mapping operation, the desired seen, reachable position will associatively evoke the required intensity of the muscle

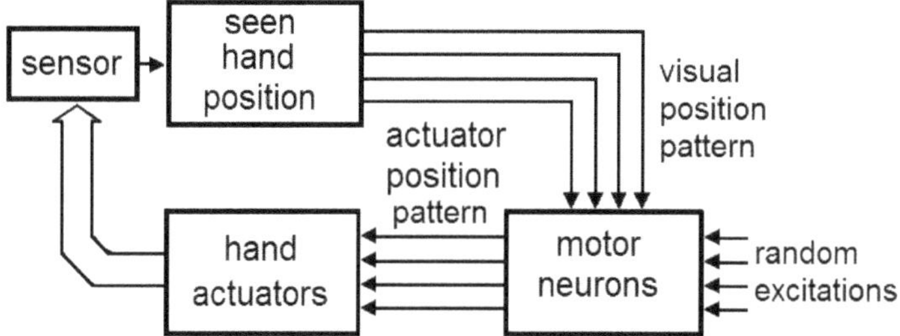

Fig. 23.1. The hand position caused by a random motor neuron excitation is seen and perceived. Position percept signals are associated with the simultaneous muscle tension command signals. Haphazard hand motions will eventually lead to the mapping of the reachable space in terms of motor neuron signals.

contraction signals. Required degrees of contractions will happen and the required hand position will follow. The required neural assembly is depicted in Fig. 23.1.

I humans, muscles do not only receive commands from the brain, they also transmit back information of their instantaneous degree of contraction, which is related to the positions of the body parts in relation to each other. Each hand position has its own unique set of the degrees of muscle contractions. The detection of these degrees of muscle contractions allows the perception of body part positions also in darkness. This applies to HCA robots as well.

23.3. Learning to Imitate

If you cannot whistle, you cannot imitate the whistling of others.

Imitation is the act of reproducing seen actions or heard sounds. There are two kinds of imitation; the instant or real-time imitation and the imitation by memory. Humans and monkeys imitate for fun, but also for learning. The ability to imitate is also important to humanoid cognitive robots, see e.g. [Reggia *et al.*, 2018].

The mainstream view in neuroscience has been that imitation is executed by special mirror neurons in the brain [Rizzolatti and Craighero, 2004]. Also, it has been presented that mirror neurons are one of the most important discoveries in the last decades of neuroscience [Acharya and Shukla, 2012; Iacoboni, 2009]. Further on, it has been theorized that the existence of these kinds of special neurons facilitate not only imitation, but also the understanding of others' minds, empathy and many other mental functions.

However, recently there have been some doubts about the validity of the mirror neuron theory and its implications. The interpretation of the results of the various research results may be inaccurate and the conclusions are subject to doubt [Spaulding, 2012].

A mirror neuron is seen as a special neuron, which is activated in two cases, namely when an action is seen and when the same action is imitated. During a seen action sequence, certain neurons are seen to alight sequentially, and during the imitation of the same sequence, the same neurons are seen to alight sequentially in the same way. Brain imaging technologies have revealed that there appears to be plenty of these kinds of neurons in the brain.

In the following it is presented that by using associative neurons, imitation can be learned and executed very simply without any need for special mirror neurons. Examples follow.

There is a fundamental precondition for all imitation; one cannot imitate something if one is not able to produce the basic elements of what is to be imitated. For example, little babies have the inborn ability to produce sounds (as all parents have noticed). Without this ability babies could not learn to speak by imitating the speech of adults.

Basically, humans produce sounds by the speech organ, which consists of the mouth, lips, the tongue and the larynx with voice cords. The speech organ controls the flow of the air exhaled through the mouth and makes the exhaled air vibrate at fundamental frequencies (pitch) and its harmonics.

The frequency of the pitch of the voice and the relative intensity of the harmonics are controlled by the various muscles of the speech organ; the production of sounds is a motor act. During imitation the speech organ must be excited in the specific way that will produce the imitated sound. This will take some learning.

The question is; how could this learning take place, considering that hearing a given sound several times over will lead to the ability to remember it, but this alone will not lead to the ability to imitate it. For example, little babies are able to hear and remember sounds, but they are not right away able to imitate them correctly.

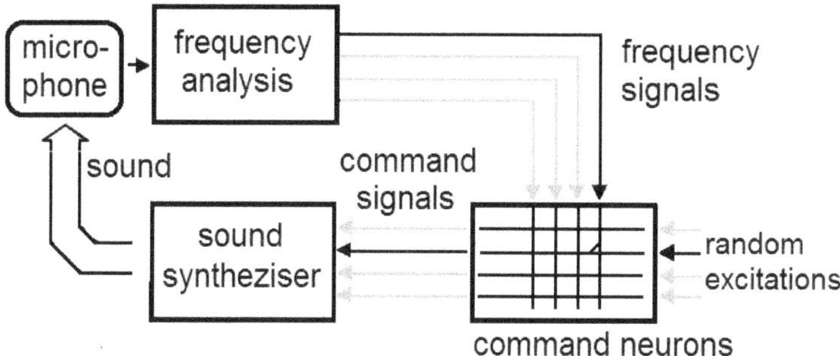

Fig. 23.2. Random excitations produce random sounds. These are detected by the microphone and their frequencies are analyzed. Ensuing frequency signals will be associated with corresponding command signals. Imitation of sounds is now possible.

The HCA supports imitation, but as in humans, also in the HCA imitation has to be learned. In HCA the learning to imitate of a single sound takes place as follows,[1] see Fig. 23.2.

The learning via random excitations works also here. In this process the command neurons of the sound synthesizer are initially excited randomly, and consequently random sounds will follow. These sounds are captured by the microphone, and the frequency analyzer will then deliver the corresponding frequency signal. Now the command neuron group will receive the random excitation signals and the percept signals at the same time, and will associate them with each other.

Later on, when the random excitation is no longer present and the same sound is heard, the sound's frequency signals will evoke exactly those command signals that will produce the heard sound.

This process can be expanded into the learning and imitation of complex sounds and sound patterns, such as spoken words.

Both the frequency signals and command signals are technically similar neural signals, and therefore one might ask, what is the reason for the use of the group of command neurons as a middle man, why not use the frequency signals as the command signals directly?

---

[1]This mechanism is experimentally verified by the author. See the "Robot mind body connection" video at https://www.youtube.com/user/PenHaiko

Indeed that would be the easiest way, but there is a problem here. The frequency signals are not compatible with the sound synthesizer command signals because they are from different modalities. Moreover, there are many frequency signals, and there are many command signals. Initially, there is no knowing which frequency signals correspond to which command signals. This is the very thing that has to be learned and has to be wired in order to become able to imitate heard sounds.

23.4.   Learning to Imitate Hand Gestures

The learning of the imitation of seen hand gestures combines the principles of the reaching out operation and the imitation of sounds.

Initially, young children execute hand gestures randomly and observe them. The hand gestures are seen as they are, but as such these are not very useful for the learning of imitation; persons to be imitated do not look the same. However, the visual perception process extracts also invariant features for the use of pattern recognition (if it were not so, we would not understand cartoons). In this case basic gesture features are extracted and are associated with the motor commands. The extracted basic gesture features are amodal; they do not depend on the external appearance or the size of the doer, be it the person to be imitated or the imitator, they depend only on the perceived patterns. As soon as the gesture features are associated with the motor commands, the seen hand gestures of other people can be imitated readily.

23.5.   Imitation, Meaning, Empathy and Mirror Neurons

According to the foregoing, associative imitation is rather mechanical and is executed without any reference to the meaning of the imitated action; only the action itself is imitated, but its purpose (if any) is not captured.

Without meaning there is no understanding. Learning to imitate some work routines will only amount to rote learning, as long as the purposes of each step are not understood. The proper learning of work routines is only then achieved, when the worker is told what the steps of the work are for and why they have to be executed in the shown way.

According to the mirror neuron theory, the ability to imitate allows the evocation of empathy. A suffering person grimaces in agony, and the imitation of this grimace will produce agony and empathy. However, grimaces can be imitated, but the imitation will not produce any feel of agony if the meaning of the grimace is not understood. But, if the observing person has suffered similar agony, then the seen grimace may evoke the feel that was associated with the observing person's grimace during the person's own agony. That may evoke empathy. If you see a little girl in tears, you might feel empathy because you know what tears mean; you have experienced the same in your childhood. No need to imitate the facial expression, but if one does, it is secondary.

This does not dovetail well with the mirror neuron theory and its proposed implications to cognition, empathy and the understanding of others' minds. Special mirror neurons are not needed for imitation, and they do not freely lead to their advertised outcomes, either. Associative neurons provide the ability to imitate, but the mere imitation will not produce the wonderful things that the mirror neuron theory posits. Meanings have to be evoked, and this would be an associative process.

23.6. The Execution of Imagined Actions

Physical interaction with the material world is more than the mere imitation of gestures; it is also the execution of planned physical actions. Technically, mental plans of actions to be executed are imaginations.

There are the imaginations of physical actions, and there are the muscles that execute these actions. Imaginations belong to the visual domain, while muscle commands do not. The question is; how to transform the imaginations into neural muscle commands?

The solution is obvious and simple. If the system were able to imitate visually perceived actions, then it would also be able to imitate visual imaginations of actions, if these were perceived as sensory percepts. In the perception-response feedback loop model the feedback loop does just that. It returns the imaginations into the perception process as virtual percepts, which then are executable in the same way as the actual sensory percepts, see Fig. 23.3.

The physical execution of imaginations must be learned. The initial random excitation principle works also here. During the initial random execution, the associative networks produce random output signals (gray lines in Fig. 23.3) that are forwarded to the motor neuron group. The signals go through this neuron group to the effectors ("muscles"), which will then

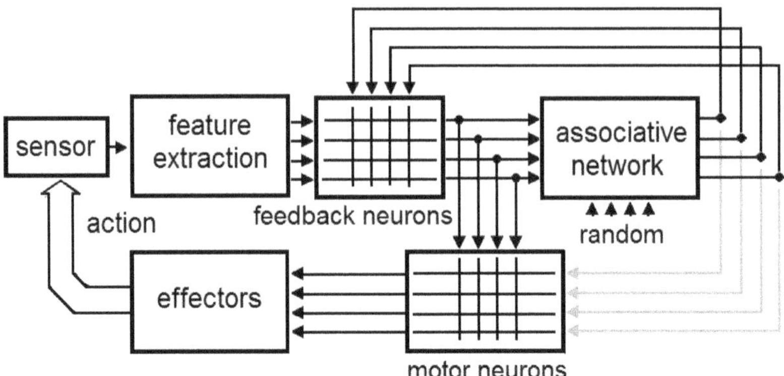

Fig. 23.3. The principle of the execution of imagined actions. The imagined actions to be executed are fed back to the perception process to be perceived as virtual percepts. These can now become reproduced as actual acts by the imitation process.

produce physical output actions. These actions are sensed, heir features are extracted and the corresponding feature signals are forwarded to the feedback neuron group. The feedback neurons receive also feedback signals from the associative network. Now, the feature signals and the feedback signals are simultaneously present and will be associated with each other. Thereafter the system "knows" which kinds of feature signals correspond to respective imagination signals. After learning, the imagination signals from the associative network will evoke corresponding virtual perception signals, which will become consciously experienced. (There is also a shortcut to this process. If the associative network is configured so that it initially allows the input signals pass, then the output signals will become automatically associated with the feature signals.)

At the same time also the motor neuron group learns the required connections, as is explained previously in the context of the imitation of sounds.

Inner speech is also imagination, while the overt spoken speech is a motor action. Therefore inner speech can be executed as overt speech by the same principle. Speaking aloud is the motor execution of imagined talk.

In functional systems, there has to be a number of controlling execution thresholds. Imitations and executions of imagined actions and speech must not be initiated automatically, they must be controlled by the needs of the actual situations. The muscle commands are there, and there is the potential of execution, but the execution cannot take place unless the commands exceed the instantaneous execution thresholds. In humans this execution control is rather subconscious.

If the neural associative principles of the HCA were applicable to the human brain, it would show that as long as there were potential connections, synapses and neurons, the brain's networks would be mainly organized not by itself, but by the simultaneously occurring sensory percepts of different sensory modalities. In these ways the various connections of the world would be captured in the brain's neural networks.

23.7.   Mind–Body Problem in the Execution of Imaginations

From immaterial imaginations to material actions. The ultimate mind–body interaction.

Imaginations are mental appearances, while the physical actions are not. Also, imaginations are just imaginations, not any real world actions, while the physical actions are very much real and material. How can they interact?

This is the old mind–body problem, and it applies to the human brain and to the robot brain equally. The solution to the mind–body problem is presented in the Chapter 3 of this book. Basically, there is no mind–body problem to begin with, as the mind is not immaterial; it only appears to be as the material processes behind it are hidden.

A conscious robot mind should also have similar immaterial appearance. That is, only the imaginations and actions should be perceived, not the material mechanisms and the connections between these.

In the HCA imaginations are generated in the visual modality, and are perceived via the feedback as virtual visual percepts in the terms of visual perception. The imaginations of the physical actions to be executed are associatively connected to the motor modality for their execution. These connections have been learned earlier, and will be activated by the imaginations of actions.

In this process the motor modality does not receive anything; the required motor elements of the imagined actions are just associatively evoked and activated. The origin of this activation is not transmitted to the motor modality, as there is no use for it and no mechanism to do it.

In the execution of imagined actions, the associative cross-connections between the visual modality and the motor modality are not perceivable to the system itself, they remain hidden and in this way they are "immaterial".

A HCA robot would be able to execute imagined actions, and for the robot there would not be any mind–body problem. Any mind–body problem here would be an external observer's misperception.

23.8.　Hierarchical Control of Motor Routines

Complicated physical actions can be imagined and executed. But there is the thing; we do not have to imagine each and every step of the imagined action. For example, when we are walking, we do not imagine and pay much attention to the act of the execution of each step that we take. We are not thinking like left foot–right foot–left foot–right foot–and so on in order to keep on walking. Walking is a routine that can be executed automatically without conscious attention, whenever we decide to walk.

Many physical actions demand continuous control and variation of the muscle command signals. That would be an excessive load to the perception-response feedback loops; they would have to concentrate only on the execution of the physical action. Each hand motion had to be continuously "supervised", each walking step would demand continuous attention and control.

Almost all of our daily physical activities are learned routines. The better we have learned them, the less we have to pay attention to their execution. We only have to decide that this or that physical task has to be initiated. And even this may not always be necessary; the situation itself may trigger the relevant action.

However, physical routines are many, and their storage and execution take actually lots of memory and processing power. In the human brain the area that is responsible for the execution of motor actions is the motor cortex area in the cerebrum. Motor cortex is involved in the learning, execution and control of voluntary movements.

But that is not all; there is another major part of the brain involved in the learning and generation of motor actions, namely the cerebellum (the "little brain" below the cerebrum).

The cerebellum is a large neural structure containing around 70 billion neurons, perhaps up to 80% of the total numbers of neurons in the brain. The cerebellum is physically smaller than the cerebrum, but its neurons are smaller. Therefore the number of its neurons can be larger than the number of neurons in the cerebrum.

The cerebellum receives information from the cerebrum and other parts of the brain. It apparently learns and contains motor routines such as walking and cycling. It delivers muscle command sequences, when the routines are commanded by the motor cortex in the cerebrum. The learned motor routines are quite permanent, maybe even lifelong.

The final link between the brain and the muscles is the spinal cord, where the final motor neurons are located. These neurons receive commands from the brain, but they can also execute fast reflexes on their own as direct responses to stimuli.

The cerebrum — motor cortex — cerebellum — spinal cord system is a kind of hierarchical command chain, where the desired action is initiated on the general level in the cerebrum, motor cortex works on details, cerebellum works on fine details and the spinal cord delivers the actual neural command signals to the muscles.

Hierarchical control of physical motions distributes processing capacity needs and allows thus the handling and planning of actions on general symbolic level. When one plans to walk to somewhere, there is no need to work out the actual motor commands, as these are produced automatically and subconsciously down in the hierarchical command chain.

In this way the workload and memory capacity requirements of the cerebrum are minimized. The cerebrum does not have to deliver the actual motor commands, and thus its mental capacity is available for more important purposes. Similar hierarchical control would be useful also in conscious robots.

Chapter 24

Precursors of Conscious Perception in the XCR-1 Robot

24.1. The XCR-1 Robot

The author's Experimental Cognitive Robot (XCR-1) is a small-scale neural hardware robot, which was designed for the practical testing of the basic principles of the HCA model. XCR-1 does not utilize program code or microprocessors. The first configuration of the XCR-1 was presented in 2010 [Haikonen, 2010, 2011], and since then its hardware neural circuitry has been revised and augmented several times. XCR-1 robot is seen in Fig. 24.1. XCR-1 is a small three-wheel differential drive robot with gripper hands. It can search for and grip small cylindrical test objects.

Fig. 24.1. The XCR-1 robot.

XCR-1 has sensory modalities for infrared vision, hearing, touch (hands) and mechanical shock (body). XCR-1 has also a "petting" sensor. All these produce non-symbolic percept signals and signal patterns.

XCR-1 operates with associations. Percepts can be used as symbols and objects can be named. In principle, the neural networks of the robot could be taught to name objects, but for practical reasons this process is pre-done; XCR-1 loses its memory, when it is switched off. Otherwise all this had to be redone before any further experimenting were possible.

XCR-1 has overt inner speech with very limited vocabulary. It has also a limited ability to recognize some spoken words by using formant detection. XCR-1 can verbally report what it is doing and what it is experiencing (me search, me touch, me hurt, me bad etc.). When asked, it can give answers that it has concluded on its own.

XCR-1 is fully autonomous, and it operates on its own without any connections with any external controlling devices.[1]

24.2. The Behaviors of the XCR-1 Robot

The basic behavior of XCR-1 is the searching for the cylindrical test objects. There are two of them, "green" and "blue". The test objects send modulated infrared radiation that the robot is able distinguish. When the robot sees one of these objects, it will home in on it and grip it and after a while it will release it and retreat. During this process XCR-1 will state "me search", "me search green" (or blue), "me touch green" (or blue). XCR-1 is also able to detect match and mismatch conditions between the sensory percepts and internally evoked percepts.

XCR-1 is able to sense mechanical vibrations when it is hit. So-called pain signal is generated, but the functional pain itself is in the form of disruptive system condition that affects attention, which in turn has secondary consequences. "Pain" causes also direct motor responses. If the robot is hit when it sees or holds one of the test objects, like the blue one, it will

---

[1] XCR-1 demo videos at https://www.youtube.com/user/PenHaiko XCR-1 early version has appeared in the Science Channel series *Through the Wormhole* by Morgan Freeman, season 4 [2013], episode 7 "Are Robots the Future of Human Evolution?".

back off hastily and make a cry and will report: "me hurt! Blue bad!". After this experience the robot will avoid the object, as badness is now associated with the object. One may now ask: "Is blue bad?" The robot will now answer: "Yes". If it is asked: "Is blue good?" the robot will answer: "No".

XCR-1 can also learn by verbal teaching. For instance, when it is first taught via petting that green is good, and then stated repeatedly that "green is bad", the robot will first deny this: "No, green good". And then again and again, but eventually the robot agrees: "Green bad". Verbal teaching is slower than direct teaching.

XCR-1 can smile and grimace as a reaction to petting, hitting and match/mismatch situations. Here the situations are real, while smile and grimace are only indicators of what is happening inside. In humans, this is rather similar; smile and grimace are only external expressions of inner mental states.

24.3. Self and Immateriality in XCR-1

The author's experimental robot XCR-1 has a rudimentary concept of self: It can verbally report "me hurt", "me search", "me touch" and it can recognize its mirror image. This concept of self is connected to the observations of the body and its states and actions.

An observer may recognize its image in the mirror, if the mirror image imitates the actions of the observer. The XCR-1 robot uses this principle to recognize itself in mirror. When the robot recognizes itself in the mirror, it will smile and report: "Me".

Consequently, the mental self-concept of the XCR-1 can be modified by putting the robot in front of a mirror and hitting the robot. The robot will now report: "Me hurt!". But then the robot will no longer smile, and it will report: "Me bad!" Thereafter the robot will avoid its mirror image. The mirror image has evoked the self concept that is now associated with badness.

The experience of the immateriality of percepts comes from the fact that the material bases of percepts cannot be observed. If the robot were able to observe and note this, then its conclusion would be: Percepts are immaterial. But the robot would not be able to reach this conclusion, if it did not have the concept of immateriality to begin with. That would not change anything, though.

Perception processes produce impressions of what is perceived. They do not produce impressions of the machinery that produces the impressions of the perceived. Therefore, percepts appear as immaterial, without any material base. (The perceived things as themselves are not taken as immaterial.) Mental content is experienced in the way of virtual percepts, therefore the above applies to it as well; thoughts, imaginations and other mental content appear as immaterial. In that sense, and only in that sense, the mind is immaterial.

A full-scale HCA-robot would be able to perceive various qualities of the sensed objects and phenomena directly in non-symbolic self-explanatory ways. The internal appearance of these qualities is immaterial, as their neural basis is not observed. These appearances of perceived qualities correspond to the human "immaterial" qualia, but their internal appearances may be, and most probably will be quite different from the human experiences of qualia.

24.4.　Is the XCR-1 Conscious?

We can ask whether the XCR-1 were conscious, and an interesting question this would be, indeed. But what exactly would we be asking?

In the previous chapters it was argued that consciousness is not an entity, it is only a vague concept. Whatever phenomena are seen as manifestations of consciousness, are explained better by other functions, such as perception and qualia.

Therefore, instead of asking about "consciousness" the following questions should be asked: Does XCR-1 have sensory *experiences*? Are these self-reportable, non-symbolic and self-explanatory? Can they be considered as forms of qualia? Does XCR-1 have the experience of being an observer, the one that observes? Does it have the experience of being the one behind its eyes? Does it have the feel of being the self?

Obviously, these questions would be most essential to all efforts in the creation of a conscious robot. If you can put somebody behind the eyes of a robot (or behind some other sensors) you will have it done. What would all this take, and does XCR-1 have what it takes?

It all begins with sensing and perception. Digital computers may utilize various sensors, such as microphones and image sensors, but these do not create any experiences of sensing anything, they produce only digital data. In humans (and animals) sensory perception is different; it creates the impression of directly seeing, hearing and observing whatever there is, without any apparent processes of coding, decoding and interpretation. The percepts are self-explanatory with qualia. Qualia are not symbols, they are directly experienced as what they are.

Thus, the first requirement is a perception process that produces direct self-explanatory percepts. XCR-1 robot has these kinds of perception processes, even though very rudimentary ones. The XCR-1 utilizes the produced percepts as such, and via association also as symbols in symbolic processing and in inner speech.

Does XCR-1 have real qualia? Qualia are subjective experiences of what is being sensed. For example, different people may see colors in different ways. Likewise, the qualia of a robot might be very different from the corresponding human qualia. However, there are amodal qualia that are universal, like rhythms.

XCR-1 can report pain and the visual detection of the test objects "green" and "blue". It could report even more, if it had higher resolution sensors. But the key word here is "the experience".

What could be considered as a sensory experience? A sensory experience is "felt", it is self-reportable and can be remembered at least for a short while. It has effects. All this is related to conscious perception.

The perception processes in the XCR-1 produce these kinds of "experiences", especially the experience of "pain". When XCR-1 is hit, it reports repeatedly that it hurts, as the condition lingers on for a while; it is actively remembered, just like the pain that one experiences after hitting one's thumb by a hammer. Also this condition is disruptive and it affects everything that the robot tries to do for a while. This is not a simulation, it really happens. But is it the real thing; is there the experiencer that has experiences, the self?

Perception process creates the experience of the self. Initially, babies sense their body by seeing it and touching it all over. This produces simultaneous touch sensations in the touching part (hand) and the touched body part; the perception of me touching me. The self-concept "me" is a product of a sensory loop, "the self-loop".

The hands of the XCR-1 robot cannot reach out and touch its body allover. The hands have touch sensors, and can sense the touching of each other, but the XCR-1 cannot see that happening. Vision and more flexible hands for touching allover would be required. The chassis of the XCR-1 does not allow this, but in principle that would be possible.

XCR-1 can report things related to "self" by using the word "me". Technically these reports are correct, but yet superficial because one function is missing: XCR-1 can report "seen" green, but it cannot *perceive itself seeing* (could be remedied), it cannot truly experience that it is the one that sees. The proper self-loop is missing.

Inner speech would be a contributor to the self-loop. XCR-1 has inner speech, and it can refer to itself by using the word "me" ("me touch"), but due to the minimal vocabulary and sensory resolution, the result is also minimal.

Additional shortcomings of XCR-1 include the lack of distinction between the external world and the robot itself. This results from the lack of the externalization and localization of percepts. That could be remedied, and if all other requirements were met, this would allow the positioning of "the me" behind the eyes of the robot.

XCR-1 does not have short-term and long-term associative memories. Thus it cannot experience time and cannot acquire personal history. This is just a technical shortcoming that could be easily remedied.

XCR-1 is not conscious, and when asked, it will state so because the heard word "conscious" does not match anything in its limited mind, as XCR-1 does not have any experienced concept of consciousness. However, it has some precursors of consciousness.

It counts what an experiment has achieved, but what has not been achieved counts even more, especially if it can be seen why something remains missing. The shortcomings of the XCR-1 result from the minimal realization, and might be remedied in more advanced systems, now that it is understood what these shortcomings are.

Chapter 25

Concluding Notes

Mind and body; one of these is touchable and concrete, the other is not. Yet, they both are or seem to be observable in living beings. The body is matter, while the mind does not appear to be material. The mind is conscious, while the material body is not. These observations have provided the intriguing premises and a good setup for philosophical contemplations for thousands of years. However, these still continuing contemplations have not led to anything concrete.

Modern physics explains everything by material particles, non-material particles (e.g. photons), forces, interactions and the concept of energy. What is observable is explainable within this framework. Modern physics has enabled our high technologies, and it serves us well. But when it comes to the phenomena of mind, self and consciousness, modern physics has appeared powerless. And indeed it is, as physical theories are mathematical models of the nature, not the mind. Physics models and explains what is observed, not what is experienced.

It has been understood that physics cannot ever explain the mind and consciousness because these are immaterial, and not observable by material instruments. Immaterial entities do not exist in the material world of physics, and therefore they cannot be explained by physical theories, either. Thus, the mystery of consciousness must be way beyond the limited capacity of human comprehension and will always remain so.

But are the mind and consciousness really immaterial entities, or entities at all? In this book the author presents that they are not; they are abstracted concepts, and moreover, very vague ones. Besides, concepts are not explained, they are defined. Accordingly, the proper explanation of consciousness would be that it is an ambiguous concept invented by humans. Concepts are concepts, not real things.

However, we are conscious and aware, when we are awake and thinking. Also, our mental content seems very much to be immaterial; there is no denying of this impression.

The impression of the immateriality of the mind has led to the mind–body problem, the problem of the interaction between the immaterial mind and the material body. This problem is an age-old philosophical one. It has not been solved so far, and probably never will be. What is immaterial, does not dovetail well with the material reality.

In this book the author presents that immateriality is just an illusion, caused by the inability to see the material machinery and processes behind the apparently immaterial phenomena. We are not able to observe our own material brain, neurons and their activities behind our mental contents, but this does not prove that the mental contents were immaterial. Hidden things are unobservable, but not necessarily immaterial. Considering this and other aspects, it can be concluded that the mind–body problem is not well-founded. It is not a real problem, it is just a philosophical misperception.

In this book the author presents that consciousness does not exist as an entity, and therefore it is a useless and misleading concept. Our experience of being conscious is created by sensory perception of the world and virtual perception of thoughts and emotions. Without percepts there is nothing to be conscious of, and we are unconscious. No "consciousness" entity is required for this.

"Consciousness" is an empty concept and the whole word should be banned. The perception processes are real, but not without a deep mystery. Our sensors produce various percepts with different qualitative appearances, also known as qualia. We are surrounded by the colorful world with its various sounds and objects. This is our qualitative impression, but in reality there are no colors, only photons with different energies and also fast air pressure vibrations. The question is; how do these produce the apparently immaterial impressions of colors and sounds, and moreover in the way that makes us to take these impressions at their face value? This is the qualia problem, also known as the hard problem of consciousness.

The explanation of phenomenal perception leads to the explanation of qualia, and in doing so it makes "consciousness" redundant. Here, it should be noted that cognition and the mental content of the mind, the "contents of consciousness", are separate and different issues, and have their own explanations, provided by psychology and cognitive sciences.

In this book the author presents that qualia are actually excitation modes of sensory neurons and neuron groups. These excitation modes are not matter; they are material reactions. These reactions are internally experienced as qualia. Here only the reactions count, not their material basis. That remains hidden, and consequently qualia appear without any kind of observable material carrier. Qualia are not observed, they are experienced. That is good, but without a problem: Who is the observer or the experiencer?

There is a difference between the outside observer and the inside experiencer. For example, as outsiders we can painlessly observe others in pain. But when we are in pain, we are inside experiencers, and the hurting pain is our inside experience; reportable though, but not an observation in the first place. The outside observer observes the external, while the inside experiencer experiences the internal. We have both roles depending on circumstances.

Qualia are self-explanatory and are taken directly what they appear to be; they are not symbols or presentations to be interpreted. This very property of qualia allows their use as foundations for cognition. Qualia percepts can be associatively connected with each other, and this facilitates the seamless transition from non-symbolic to symbolic processing and the generation of natural languages and inner speech — and it also facilitates the ability to create concepts about the world. What does all this amount to, if not to everything?

The viewpoint matters. The author presents that the mysteries of mind–body interactions, immaterialism, consciousness and qualia can be dissolved by the power of an altered viewpoint, one that allows the peeking behind the age-old concepts. When illusions, misperceptions and false preconceptions are swept away, also mysteries disappear.

References

Aleksander, I. [2009]. "The potential impact of machine consciousness in science and engineering". *International Journal of Machine Consciousness*, 1(1), 1–9.

Aleksander, I. [2015]. *Impossible Minds, My Neurons, My Consciousness*, revised ed. (UK: Imperial College Press).

Acharya, S., Shukla, S. [2012]. "Mirror neurons: Enigma of the metaphysical modular brain". *Journal of Natural Science, Biology and Medicine*, 3(2), 118–124.

Baars, B. J. [1997]. *In the Theater of Consciousness* (Oxford: Oxford University Press).

Baars, B. J., Franklin, S. [2009]. "Consciousness is computational: The LIDA model of global workspace theory". *IJMC*, 1(1), 23–32.

Barbour, J. [1999]. *The End of Time* (Oxford: Oxford University Press).

Bołtuć, P. [2020]. "Consciousness for AGI". *Procedia Computer Science*, 169, 365–372.

Bostrom, N. [2003]. "Are you living in a computer simulation?" *Philosophical Quarterly*, 53(211), 243–255.

Brian, E. [2014]. "The biological function of consciousness". *Frontiers in Psychology*, 5, 697.

Chalmers, D. J. [1995]. "Facing up to the problem of consciousness". *Journal of Consciousness Studies*, 2(3), 200–219.

Chalmers, D. J. [1996]. *The Conscious Mind* (Oxford: Oxford University Press).

Chella, A. [2008]. "Perception loop and machine consciousness". *APA Newsletter on Philosophy and Computers*, 8(1), 7–9.

Church, A. [1936]. "An unsolvable problem of elementary number theory". *American Journal of Mathematics*, 58, 345–363.

Crick, F. C., Koch, Ch. [2005]. "What is the function of the claustrum?". *Philosophical Transactions of the Royal Society B*, 360(1458), 1271–1279.

Damasio, A. [2000]. *The Feeling of What Happens* (UK: Vintage).

Davenport, D. [2012]. "Computationalism: Still the only game in town". *Minds and Machines*, 22(3), 183–190.

Dennett, D. [1991]. *Consciousness Explained* (USA: Little, Brown and Company).

Dijkstra, N. [2024]. "Uncovering the role of the early visual cortex in visual mental imagery". *Vision (Basel)*, 8(2), 29.

Goldstein, E. B. [2002]. *Sensation and Perception*, 6th ed. (USA: Wadsforth).

Goodman, N. [1977]. "Topology of quality". In: *The Structure of Appearance. Boston Studies in the Philosophy of Science*, 53 (DE: Springer).

Eagleman, D. *et al.* [2005]. "Time and the brain: How subjective time relates to neural time". *The Journal of Neuroscience*, 25(45), 10369–10371.

Haikonen, P. O. [1999]. *An Artificial Cognitive Neural System Based on a Novel Neuron Structure and a Reentrant Modular Architecture with Implications to Machine Consciousness*. Doctoral Thesis. Series B: Research Reports B4. Helsinki University of Technology, Applied Electronics Laboratory.

Haikonen, P. O. [2003]. *The Cognitive Approach to Conscious Machines* (UK: Imprint Academic).

Haikonen, P. O. [2007]. *Robot Brains; Circuits and Systems for Conscious Machines* (UK: John Wiley & Sons).

Haikonen, P. O. [2010]. "An experimental cognitive robot", in A. V. Samsonovich, K. R. Johannsdottir, A. Chella and B. Goertzel (eds.), *Biologically Inspired Cognitive Architectures 2010* (Amsterdam: IOS Press), 52–57.

Haikonen, P. O. [2011]. XCR-1: "An experimental cognitive robot based on an associative neural architecture". *Cognitive Computation*, 3(2), 360–366.

Haikonen, P. O. [2012]. *Consciousness and Robot Sentience* (Singapore: World Scientific).

Haikonen, P. O. [2019]. *Consciousness and Robot Sentience*, second ed. (Singapore: World Scientific).

Haikonen, P. O. [2023]. *Existence, Origin and Weird Technology* (Singapore: World Scientific).

Harada, K., Kamiya, T., Tsuboi, T. [2016]. "Gliotransmitter release from astrocytes: Functional, developmental, and pathological implications in the brain". *Frontiers in Neuroscience*, 9, 499.

Hebb, D. O. [1949]. *The Organization of Behavior* (New York: Wiley).

Hesslow, G. [2002]. "Conscious thought as simulation of behaviour and perception". *Trends in Cognitive Sciences*, 6(6), 242–247.

Hurlburt, R. T. [2011]. "Thinking without words or images is possible". *Psychology Today*. https://www.psychologytoday.com/us/blog/pristine-inner-experience/201111/thinking-without-words.

Iacoboni, M. [2009]. "Imitation, empathy, and mirror neurons". *Annual Review of Psychology*, 60, 653–670.

Jackson, F. [1982]. "Epiphenomenal qualia". *Philosophical Quarterly*, 32, 127–136.

James, W. [1890]. *The Principles of Psychology* (USA: Harvard University Press).

Johnson, K. O. [2000]. "Neural coding". *Neuron*, 26(3), 563–566.

Kinouchi, Y., Mackin, K. J. [2018]. "A basic architecture of an autonomous adaptive system with conscious-like function for a humanoid robot". *Frontiers in Robotics and AI*. https://www.frontiersin.org/journals/robotics-andai/articles/10.3389/frobt.2018.00030/full.

LeDoux, J. [1996]. *The Emotional Brain* (New York: Simon & Schuster).

Lefton, K. *et al.* [2025]. "Norepinephrine signals through astrocytes to modulate synapses". *Science*, 388(674), 776–783.

Li, M., Tsien, J. [2017]. "Neural code—neural self-information theory on how cell-assembly code rises from spike time and neuronal variability". *Frontiers in Cellular Neuroscience*, 11, 236.

Makin, S. [2024]. "Not everyone has an inner voice streaming through their head". *Scientific American*, https://www.scientificamerican.com/article/not-everyone-has-an-inner-voice-streaming-through-their-head/.

McCulloch, W. S., Pitts, W. [1943]. "A logical calculus of ideas immanent in nervous activity". *Bulletin of Mathematical Biophysics*, 5, 115–133.

Miller, E. K., Cohen, J. D. [2001]. "An integrative theory of prefrontal cortex function". *Annual Review of Neuroscience*, 24(1), 167–202.

Morin, A., Everett, J. [1990]. "Inner speech as a mediator of self-awareness, self-consciousness, and self-knowledge: An hypothesis". *New Ideas in Psychology*, 8(3), 337–356.

Nedergaard, J. S., Lupyan, G. [2024]. "Not everybody has an inner voice: Behavioral consequences of anendophasia". *Psychological Science*, 35(7), 780–797.

Nichols, J., Martin, A., Wallace, B. [1992]. *From Neuron to Brain*, 3rd ed. (USA: Sinauer Associates, Inc.).

Perea, G., Navarrete, M., Araque, A. [2009]. "Tripartite synapses: Astrocytes process and control synaptic information". *Trends in Neurosciences*, 32(8), 421–431.

Reggia, J. A., Katz, G. E., Davis, G. P. [2018]. "Humanoid cognitive robots that learn by imitating: Implications for consciousness studies". *Frontiers in Robotics and AI*, 5, 1.

Rizzolatti, G., Craighero, L. [2004]. "The mirror-neuron system". *Annual Review of Neuroscience*, 27(1), 169–192.

Rosenblatt, F. [1958]. "The perceptron: A probabilistic model for information storage and organization in the brain". *Psychological Review*, 65(6), 386–408.

Samsonovich, A. V. [2010]. "Toward a unified catalog of implemented cognitive architectures", in A. V. Samsonovich, K. R. Johannsdottir, A. Chella and B. Goertzel (eds.), *Biologically Inspired Cognitive Architectures 2010* (Amsterdam: IOS Press), 195–244.

Sanz, R. *et al.* [2009]. "Systems, models and self-awareness: Towards architectural models of consciousness". *International Journal of Machine Consciousness*, 1(2), 255–279.

Seth, A. [2021]. *Being You: A New Science of Consciousness* (New York: Faber/Dutton).

Scemes, E., Giaume, Ch. [2006]. "Astrocyte calcium waves". *Glia*, 54(7), 716–725.

Shanahan, M. [2010]. *Embodiment and the Inner Life* (Oxford: Oxford University Press).

Spaulding, S. [2012]. "Mirror neurons are not evidence for the simulation theory". *Synthese*, 189, 515–534.

Spiegelberg, H. [1961]. "On the "I-am-me" experience in childhood and adolescence". *Psychologia*, 4, 135–146. Lawrence College, Appleton, Wisconsin

Squire, L. [2009]. "The legacy of patient H.M. for neuroscience". *Neuroview*, 61(1), 6–9.

Steels, L. [2003]. Language re-entrance and the "inner voice", in O. Holland (ed.), *Machine Consciousness* (UK: Imprint Academic), 173–185.

Takeno, J., Inaba, K., Suzuki, T. [2005]. "Experiments and examination of mirror image cognition using a small robot", in *Proc. 6th IEEE International Symposium on Computational Intelligence in Robotics and Automation (CIRA 2005)*, 493–498.

Thompson, R. [1993]. *The Brain* (New York: W. H. Freeman and Company).

Tononi, G. [2004]. "An information integration theory of consciousness". *BMC Neuroscience*, 5, 42. DOI:10.1186/1471-2202-5-42.

Tononi, G. [2008]. "Consciousness as integrated information: A provisional manifesto". *Biological Bulletin*, 215(3), 216–142.

Tononi, G., Edelman, G. M., Sporns, O. [1998]. "Complexity and coherency: Integrating information in the brain". *Trends in Cognitive Sciences*, 2(12), 474–484.

Vaaga, C., Borisovska, M., Westbrook, G. [2014]. "Dual-transmitter neurons: Functional implications of co-release and co-transmission". *Current Opinion in Neurobiology*, 29, 25–32. Elsevier.

Index